Mass Spectrometry
and NMR Spectroscopy
in Pesticide Chemistry

Environmental Science Research

Editorial Board

Alexander Hollaender
Biology Division
Oak Ridge National Laboratory
Oak Ridge, Tennessee
and
University of Tennessee, Knoxville

Ronald F. Probstein
Department of Mechanical Engineering
Massachusetts Institute of Technology
Cambridge, Massachusetts

E. S. Starkman
General Motors Environmental Activities Staff
General Motors Technical Center
Warren, Michigan

Bruce L. Welch
Director, Environmental Neurobiology
Friends Medical Science Research Center, Inc.
and
Department of Psychiatry & Behavioral Sciences
The Johns Hopkins University School of Medicine
Baltimore, Maryland

Mass Spectrometry and NMR Spectroscopy in Pesticide Chemistry

Edited by

Rizwanul Haque

*Department of Agricultural Chemistry
and Environmental Health Sciences Center
Oregon State University
Corvallis, Oregon*

and

Francis J. Biros

*U.S. Environmental Protection Agency
Primate and Pesticides Effects Laboratory
Perrine, Florida*

PLENUM PRESS • NEW YORK-LONDON

Library of Congress Caloioging in Publication Data

Main entry under title:

Mass spectrometry and NMR spectroscopy in pesticide chemistry.

(Environmental science research, v. 4)
"Proceedings of a symposium under the auspices of the Division of Pesticide
Chemistry held during the 165th national meeting of the American Chemical
Society in Dallas, Texas, April 9-13, 1973."
Includes bibliographical references.
1. Pesticides—Analysis—Congresses. 2. Mass spectrometry—Congresses. 3.
Nuclear magnetic resonance spectroscopy—Congresses. I. Haque, Rizwanul, ed.
II. Biros, Francis J., 1938-   ed. III. American Chemical Society. Division of
Pesticide Chemistry. [DNLM: 1. Nuclear magnetic resonance—Congresses. 2.
Pesticides—Analysis—Congresses. 3. Spectrum analysis—Congresses. W1
EN986F v. 4 1973/WA240 M414 1973]
SB951.M512   668'.65   73-20005
ISBN 0-306-36304-6

Proceedings of a symposium held under the auspices of the Division of
Pesticide Chemistry during the 165th National Meeting of the American Chemical
Society in Dallas, Texas, April 9-13, 1973

© 1974 Plenum Press, New York
A Division of Plenum Publishing Corporation
227 West 17th Street, New York, N.Y. 10011

United Kingdom edition published by Plenum Press, London
A Division of Plenum Publishing Company, Ltd.
Davis House (4th Floor), 8 Scrubs Lane, Harlesden, London, NW10 6SE, England

PREFACE

Organic and inorganic chemicals employed as pesticides have achieved great economic and commercial importance in agriculture and related fields as useful agents protecting plants and animals alike. Pesticides have been immensely beneficial in controlling plant disease, insect infestation and weed growth, as well as increasing agricultural crop production all over the world. However, until recently, only a token amount of attention has been given to certain aspects of the chemical behavior of pesticides and particularly their distribution, alteration, and persistence in the environment, and their biological mechanism of action. Increasing public awareness of real and potential danger to the environment as well as the public health has resulted in greater efforts to fathom the chemical behavior of pesticides. Scientists from many disciplines are combining their efforts to gain knowledge of the mechanism of pesticide action and to solve the problems arising due to their encroachment on the environment.

In this new field of "Pesticide Chemistry" which has emerged, modern analytical instrumentation has played a significant role. Invaluable instrumental techniques have included mass spectrometry and nuclear magnetic resonance spectroscopy. Each has been used successfully in determining the concentration of pesticides in human, animal, and environmental substrates, elucidating chemical structure of intact residues and metabolic and chemical degradation products and providing insight as to their biochemical mode of action. To review these studies, examine the current state of the art, and explore the full potential of these two powerful techniques, a symposium was planned under the auspices of the Division of Pesticide Chemistry of the American Chemical Society.

This volume is a compendium of the papers presented at the symposium held during the 165th National Meeting of the American Chemical Society in Dallas, Texas, April, 1973. The contributors are all expert investigators employing either or both techniques to study some aspect of pesticide chemistry or biochemistry. The chapters represent a cross section of research performed at laboratories in the United States, Canada, and the United Kingdom from industry, government, and academia.

v

The volume is divided into two parts; the first is devoted to mass spectrometric studies and the second describes research employing the nuclear magnetic resonance technique. The mass spectrometry section includes papers detailing the use of low and high resolution techniques, individually and in combination with gas chromatography in elucidating the structure of pesticides and their degradation products, innovative approaches employed in the application of mass spectrometry to pesticide confirmation and characterization including ion kinetic energy spectroscopy, direct analysis of daughter ions, positive and negative ion chemical ionization mass spectrometry, and mass fragmentography, and finally, computer data acquisition and processing techniques for greater speed, flexibility, and accuracy in structural assignments made on the basis of mass spectral data.

The nuclear magnetic resonance section contains papers concerned with the use of this technique in: 1) structural elucidation, 2) the study of the interaction of pesticides with phospholipids, model membranes, and proteins, and 3) studies in the adsorbed state. In addition, there are chapters on the Fourier Transform technique and its application to nuclear magnetic resonance and infrared spectroscopy. A description of the application of spin-labelling to the study of pesticide interactions in vivo employing electron spin resonance methods has also been included.

This volume encompasses a broad range of topics related to pesticide chemistry. Each author should be credited individually for his contribution. If the contents of the book have in any way been successful in exploring the full potential of the two analytical techniques in solving problems concerned with the behavior of pesticides and related chemicals in the environment, then we feel our goal has been accomplished. We wish to thank all the contributors for their efforts and cooperation, without which this volume would not have been possilbe. We are indebted to several publishers who graciously permitted the reproduction or reprinting of copyright material, including Academic Press (Molecular Pharmacology); American Association for the Advancement of Science (Science); American Chemical Society (Biochemistry, Journal of American Chemical Society); American Society of Biological Chemists (Journal of Biological Chemistry); Elsievier Publishing Company (Advances in Molecular Relaxation Processes, Science of Total Environment); Heydon and Son, Ltd. (Organic Mass Spectrometry); and National Research Council of Canada (Canadian Journal of Chemistry).

Our special thanks are due to the Division of Pesticide Chemistry of the American Chemical Society, and its officers and membership for sponsoring the symposium on mass spectrometry and nuclear magnetic resonance spectroscopy in pesticide chemistry and contributing to its success.

Rizwanul Haque
Department of Agricultural Chemistry
and Environmental Health Sciences
    Center
Oregon State University
Corvallis, Oregon 97331

Francis J. Biros
U.S. Environmental Protection Agency
Primate & Pesticides Effects
    Laboratory
Perrine, Florida 33157

CONTENTS

MASS SPECTROMETRY

MASS SPECTROMETRY

MASS SPECTROMETRY IN THE STUDY OF PESTICIDES: AN INTRODUCTION

F. J. Biros

Primate & Pesticides Effects Laboratory
U. S. Environmental Protection Agency
Perrine, Florida 33157

During the preceding ten years, mass spectrometry has enjoyed a phenomenal growth as a technique which provides identification and elucidation of structure of biologically active compounds. Biochemical applications have proliferated intensely and it is rare to read a technical publication dealing with a study of the metabolism or degradation of biologically significant compounds without a reference to the use of mass spectrometry in the characterization of compounds of unknown structure. To a similar extent, mass spectrometry has been increasingly applied in the area of pesticide chemistry as an analytical technique for the confirmation of pesticide residues in environmental substrates and the characterization of intact and degraded residues of unknown structure. From the earliest recognition of the utility of this valuable technique in evaluating metabolic pathways[1], to the most recent sophisticated application to the environmental monitoring of 2,3,7,8-tetrachlorodibenzo-p-dioxin[2], mass spectrometry has continually proven to be an indispensable tool in solving the wide variety of problems presented by research and development activity in pesticide chemistry.

Mass spectrometry is routinely employed in studies of the environmental degradation, i.e., chemical and photochemical alteration, of pesticides and other toxic substances; qualitative and quantitative monitoring of residues in human, animal, and environmental media; pesticide mechanism of action studies, e.g. involving use of stable isotopes; providing supporting identification and confirmation in method development studies involving other analytical techniques such as gas liquid chromatography, liquid chromatography, and thin layer chromatography; and finally in characterizing insect and mammalian metabolites in biochemical studies. Availability of more sensitive instrumentation, providing mass spectral data at the

3

picogram level under proper conditions, and the lower cost and complexity and greater reliability and dependability of modern equipment has assured virtually every researcher in pesticide chemistry access to the mass spectrometric technique. Furthermore, the use of computers in acquiring and processing mass spectral data is becoming more commonplace. In many applications, such as gas chromatographic-mass spectrometric analysis of complex mixtures of organic pollutants in plant effluents, computers are recognized as an integral and necessary part of the basic mass spectrometer analytical system providing data at higher speed and accuracy than possible with manual processing.

In the following section, present research emphasis in the application of electron impact mass spectral techniques to the characterization of several classes of pesticides, their metabolites, and chemical and photochemical degradation products is described. In addition, pioneering applications to pesticide research of recently developed innovative mass spectral techniques are presented. These include studies of the utility and application of positive and negative ion chemical ionization mass spectrometry in the characterization of pesticidal compounds and residues of industrial chemicals in the environment. The potential utility of metastable transitions and ion kinetic energy spectroscopy in the elucidation of structure of individual pesticides and isomeric groups of pesticidal compounds is also explored. Finally, the current state of the art in the application of computer technology to the acquisition and processing of mass spectral information in pesticide research and monitoring, and the computer assisted confirmation, characterization, and elucidation of structure of pesticide molecules and related compounds is examined.

REFERENCES

1. F. A. Gunther, Instrumentation in Pesticide Residue Determination, Adv. Pest Control Research, 5, 191 (1962).

2. R. W. Baughman and M. S. Meselson, An Analytical Method for Detecting Dioxin, Abstracts, N.I.E.H.S. Conference on Chlorinated Dibenzodioxins and Dibenzofurans, National Institute of Environmental Health Sciences, Research Triangle Park, N. C., April 1973.

CHARACTERIZATION OF CHLORINATED HYDROCARBON PESTICIDE METABOLITES

OF THE DDT AND POLYCYCLODIENE TYPES BY ELECTRON IMPACT AND CHEMICAL

IONIZATION MASS SPECTROMETRY

J. D. McKinney, E. O. Oswald, S. M. dePaul Palaszek,

and B. J. Corbett

National Institute of Environmental Health Sciences

P. O. Box 12233, Research Triangle Park, N. C. 27709

Abstract. The electron impact and chemical ionization mass spectra for essentially all of the known metabolites, and a few proposed metabolites, and their derivatives of the chlorinated hydrocarbon pesticides, DDT, dieldrin, and aldrin (and their isomers) have been examined. The EI and CI spectra of these systems are in most cases qualitatively and quantitatively different, and, therefore, chemical ionization mass spectrometry adds a new dimension to the MS identification and characterization of these and related compounds as potential environmental agents. In addition, the chemical information derived from both the EI and CI spectra is appreciable and valuable in studying the metabolism and other degradative processes of these compounds such as their photolysis and thermolysis. Not only was general chemical information obtainable from the mass spectra, but it was also possible to derive stereochemical information. A comprehensive study of this type has clearly demonstrated the combined advantages of EI and CI mass spectrometry in differentiating closely related systems. Minimum detectability was not a major concern in these studies, but with the aid of multiple ion detection in conjunction with the qualitatively different EI and CI mass spectral patterns for these compounds, it should be possible to confidently identify amounts in the low nanogram or even in the picogram range.

Mass spectrometry (MS) has been used somewhat more extensively than nmr in the area of pesticide chemistry, especially as an analytical tool[1-4]; however, its usage is still small relative to that in other areas of chemistry as evidenced by the numerous publications on the subject in these other areas. The chlorinated hydrocarbon pesticides of the DDT and polycyclodiene type have had some mass spectral elaboration in the literature;[5] however, it has been largely done with electron impact systems. Chemical ionization systems[6-7] among others are receiving expanding use in all areas of chemistry, including pesticides. Such alternative approaches to the ionization of molecules and subsequent measurement of ions are being sought with the aim of simplifying mass spectral interpretation as well as gaining a more panoramic view of the entire ionization picture which should include both positive and negative ions[8-9] as well as radical ions. Chemical ionization mass spectrometry has the advantages of being a secondary ionization process in which sample molecules undergo ion-molecule reactions with a reagent gas (generally methane) to form even-electron (generally positive ions). Such a process is accomplished at lower energies and forms more stable ions and, therefore, simplifies spectral interpretation considerably.

Our laboratory is in a somewhat unique position of having both electron impact (EI) and chemical ionization (CI) systems from the same company which can be interchangeably interfaced to one computer system for collecting and processing of data. This enables us to more confidently make comparison statements regarding the advantages and disadvantages of one system over the other with a given series of compounds and other things being equal. With this in mind, both the EI and CI spectra of the chlorinated hydrocarbon pesticide metabolites of the DDT and polycyclodiene type have been examined with the hope of adding a new dimension to the MS identification of these systems as potential environmental agents. In addition, even though only two classes of compounds were examined, the functional types of compounds examined are many and include derivatives normally used in facilitating the analysis of highly polar oxidative as well as reductive metabolites. Therefore, the information obtained may have general importance in the MS analysis of related compounds in different classes.

As one author has put it,[10] mass spectra present a "chemical" appearance of compounds. Therefore, in addition to the analytical value of the data obtained, information was sought which would reflect the chemistry and, in particular, the stereochemistry of these systems and its possible translation into interactions and further metabolism in biological systems. In addition, it seemed probable that differences in the mass spectra of some of the isomeric polycyclodiene systems might reflect steric differences in the molecules.

The unimolecular ion reactions in the mass spectrometer are in many ways similar to the reactions taking place under thermal and photolytic conditions. Therefore, it was of interest to utilize the mass spectrometer as a model system to measure the decomposition and rearrangement propensities of certain of these chlorinated hydrocarbon pesticide systems. Such information would be particularly helpful to the pesticide residue chemist who must contend with thermally and photolytically degradable pesticides.

Experimental Section

Standards. The following DDT compounds (structures shown in Fig. 1) were commercially available: 4,4'-Dichlorobenzophenone (DBP), 4,4'-Dichlorobenzhydrol (DBH), Bis(p-chlorophenyl)methane (DDM), 1,1-Bis(p-chlorophenyl)-2,2,2-trifluoroethane (DDT-F), 1,1-Bis(p-chlorophenyl)-2,2,2-trichloroethanol (Kelthane), 1,1-Bis(p-chlorophenyl)-2,2-dichloroethene (DDE), 1-(o-Chlorophenyl)-1-(p-chlorophenyl)-2,2,2-trichloroethane (o,p'-DDT), and 1,1-Bis(p-chlorophenyl)-2,2-dichloroethane (DDD). The remaining DDT compounds examined were synthesized according to their published procedures[11-12] or slight modifications thereof. These compounds include 1,1-Bis(p-chlorophenyl)-2-chloroethene (DDMU), 1,1-Bis(p-chlorophenyl)-2-chloroethane (DDMS), 2,2-Bis(p-chlorophenyl)ethanol (DDOH), 1,1-Bis(p-chlorophenyl)-2,2,2-trichloroethyl acetate (Kelthane acetate), 1,1-Bis(p-chlorophenyl)-ethene (DDNU), 1,1-Bis(p-chlorophenyl)-2-ethyl acetate (DDOH-acetate), ethyl-2,2-bis(p-chlorophenyl)-acetate (DDA-ethyl ester), 2,2-Bis(p-chlorophenyl)acetaldehyde (DOCHO), 1,1-Bis(p-chlorophenyl)-2-ethenyl acetate (DDCHO-enol ether acetate), and 1,1-Bis(p-chlorophenyl)-3-oxa-1-pentenyl-5-acetate (EEAC).

The following polycyclodiene compounds (structures shown in Figs. 12-13) were commercially available: Aldrin (HHDN), dieldrin (HEOD), endrin and delta-ketoendrin. The remaining polycyclodiene compounds were synthesized according to their published procedures[13] or slight modifications[14] of methods used to prepare similar compounds, except for C-12 hydroxy dieldrin-F-1, which was obtained from metabolism of dieldrin[15] (in rat). These compounds include photodieldrin (PD), 4-oxo-4,5-dihydroaldrin (aldrin ketone-AK), exo- and endo-4-hydroxy-4,5-dihydroaldrin (xAA and nAA), exo-4-hydroxy-5-oxo-4,5-dihydroaldrin (aldrin hydroxyketone-AHK and its ene-diol tautomer), cis and trans-4,5-dihydroxy-4,5-dihydroaldrin (CAD and TAD), and the corresponding silyl ether derivatives of all these hydroxylic compounds as well as the diacetates of CAD and TAD. The majority of the polycyclodiene systems examined have well defined stereochemical structures.[13]

MS Analysis. For GC-MS analysis three columns were employed
for all of the compounds examined. These columns were a 3% OV-1
column, a one to one mixed column of 10% OV-225 and 3% QF-1, and a
11% OV-17 + QF-1 column, all of which contained 80/100 Gas Chrom Q
as support phase. Both the electron impact (all spectra taken at
70 ev) and the chemical ionization systems consisted of Varian
Aerograph Model 1400 gas chromatographs interfaced to Finnigan
Model 1015 (E&C) Quadrupole mass spectrometers. For EI-MS analysis
helium was used as carrier gas at a flow rate of 35-40 ml per minute.
For CI-MS analysis methane and isobutane were used as carrier and
reagent gas with the flow rate adjusted to maintain 600 microns of
pressure in the ionizer chamber. Best results were obtained with
the CI-MS system when the ionizer temperature was maintained between 50-75°C by manifold heat transfer.

Specific column conditions and procedures for analysis of the
DDT compounds were as follows. For EI-MS analysis of the mixed
standard of DDT compounds (See Fig. 2A-B) on OV-17 + QF-1, 0.5
microliter of the acetone solution (containing approximately
1 microgram of each component) was injected onto the column at 160°C
with the ionizer off; the column temperature program from 160-225°C
at a rate of 10°/minute was initiated immediately following injection. After 6.75 minutes the ionizer was turned on simultaneously
with initiation of the computer control for collection of data
(System 150 data collection system). In a similar manner, the CI-MS
analysis was carried out on the same column, but only 3 minutes
elapsed prior to turning on the ionizer and initiating computer
control. The special mixture of DDT compounds (DBP, DDE, DDT, KOAc)
was analyzed on 3% OV-1 in a similar fashion using 0.5 microliter
of the solution for injection with the same program. The ionizer
and computer control were turned on 4.5 minutes following injection
for EI and 1.5 minutes for CI. DDCHO and EEAC (Fig. 10) were also
analyzed with the 3% OV-1 column under the same temperature pro-
gramming conditions. The mixture of DDT and its trifluoro analog
(F-DDT) were analyzed on the OV-17 + QF-1 column under the same
program conditions (See Fig. 7). Other spectra were obtained by the
direct probe inlet including o,p'-DDT.

The mixed standard of the polycyclodiene systems (included
aldrin, dieldrin, aldrin ketone mixed with the silyl ether derivatives of the aldrin alcohols (xAA + nAA) and the aldrin diols (TAD
and CAD) was analyzed on both the OV-225 + QF-1 column and the OV-1
column since separation of all components was not achieved on one
column. A portion of the benzene solution (0.5 microliter, consisting of one microgram or less of each component) was injected onto
the OV-225 + QF-1 column at 160°C with the ionizer off; the column
temperature program from 160-225°C at a rate of 10°/minute was
initiated. After 4 minutes the ionizer was turned on together with
initiation of the computer control for collection of the EI-MS data.
For collection of data with the OV-1 column, the temperature program

was 150°-240°C at a rate of 4°/minute with ionizer and computer initiation at the end of 4.5 minutes. In similar fashion the CI-MS data was collected for those compounds on the same columns using the same temperature program and somewhat shorter turn on and initiation periods. The CI and EI mass spectra obtained for the remaining polycyclodiene systems examined were obtained either by the direct probe inlet into the mass spectrometer or GC on the OV-17 + QF-1 or OV-1 columns under isothermal conditions (225°C). Endrin, photo-dieldrin, C-12 hydroxydieldrin (metabolite-F-1), delta-ketoendrin, and the aldrin hydroxy ketone mixture (see Fig. 20) were chromato-graphed separately on 3% OV-1 at 225°C. The products derived from chemical derivatization of dieldrin according to the Chau-Cochrane method[16] were chromatographed on the 11% QF-1 + OV-17 column at 225°C (See Fig. 21).

Discussion

The discussion of the mass spectra of these compounds will be divided into two groups, the DDT and polycyclodiene types, which will be further divided into their EI and CI mass spectra. Various stereochemical and other considerations suggested by the data will be discussed as they become appropriate. The DDT compounds examined are listed in Fig. 1 according to their abbreviated title, molecular formula, and molecular weight based on Cl^{35}. All the mass spectral ions discussed refer to the Cl^{35} ion unless indicated otherwise. Figure 2A shows the computer reconstructed gas chromatogram [(RGC)- see Experimental section for a discussion of the instrumentation procedures and chromatographic conditions] of a mixed standard of thirteen DDT compounds, all of which have in common the bis(p-chlorophenyl) grouping and occur as metabolites of p,p -DDT in various biological systems. Two of the thirteen are chemical deri-vatives, viz., the ethyl ester of DDA and the acetate of DDOH which possess the same empirical formula. DDOH also appears in its un-derivatized form in this chromatogram. It should be noted that kelthane and the 4,4'-dichlorobenzophenone (DBP) do not, in fact, have the same chromatographic behavior on this column, but DBP is an on-column decomposition product of kelthane. Attempts to obtain MS spectra of kelthane via the direct probe (DP) route also failed since the spectra obtained were identical to those of DBP. On the other hand, DDMU and DBH do have essentially identical chromatograph-ic behavior on this column since computer limited mass searches (LMS) confirm the presence of both with DDMU running slightly ahead of DBH. Nevertheless, this is a good example of the separation properties for a complex mixture of related compounds on the gas chromatograph interfaced to the EI-MS system.

These DDT compounds are further divided into three subgroups, viz., the saturated types including DDT, DDD, DDMS, and DDM; the unsaturated types including DDE, DDMU, and DDNU; and the oxidative

Fig. 1. DDT compounds.

DDM	DDMU	DDE
$C_{13}H_{10}Cl_2$	$C_{14}H_9Cl_3$	$C_{14}H_8Cl_4$
236	282	316

DDNU	DDMS	DDD
$C_{14}H_{10}Cl_2$	$C_{14}H_{11}Cl_3$	$C_{14}H_{10}Cl_4$
248	284	318

DBP	DDT-F	DDT
$C_{13}H_8OCl_2$	$C_{14}H_9Cl_2F_3$	$C_{14}H_9Cl_5$
250	304	352

DBH	DDCHO-enol ether acetate	o,p'-DDT
$C_{13}H_{10}OCl_2$	$C_{16}H_{12}O_2Cl_2$	$C_{14}H_9Cl_5$
252	306	352

DDCHO	DDA-ethyl ester	Kelthane
$C_{14}H_{10}OCl_2$	$C_{16}H_{14}O_2Cl_2$	$C_{14}H_9OCl_5$
264	308	368

DDOH	DDOH-acetate	Kelthane-acetate
$C_{14}H_{12}OCl_2$	$C_{16}H_{14}O_2Cl_2$	$C_{16}H_{11}O_2Cl_5$
266	308	410

R =

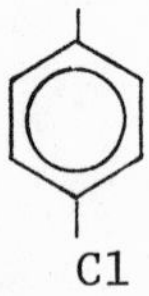

Abbrev. Title,
Formula, Molecular wt.
based on Cl^{35}

types including DBH, DBP, DDOH, DDOH-acetate, and DDA ethyl ester. The mass spectra (EI-70ev) of the saturated metabolites have several fragments in common corresponding to the loss of HCl, Cl singly or in combination as well as the facile cleavage of the C-C bond of the ethane portion of the molecule. This latter process results in a fragment ($C_{13}H_9Cl_2$) of m/e 235 which appears as the base peak in all of the saturated systems except DDM where m/e 236 is the parent peak (≈56%) and 201 (M-Cl) is the base peak. It is interesting that none of these systems except DDM (≈17%) show a fragment corresponding to the loss of the p-chlorophenyl group in their normalized spectra. The parent ions for DDT and DDD are very weak (<1%) with some improvement (≈4%) in DDMS. This probably reflects the steric requirements of the chlorine substituents on the β-carbon which are relieved with the loss of chlorine atoms. Steric hindrance to rotation about the C-C bonds is obvious on examination of space filled models of DDT.

The unsaturated compounds DDE, DDMU, and DDNU have in common the loss of Cl_2 (presumably the phenyl bound chlorines) to form the fragments m/e 246, 212, and 178 found as the base peaks in their spectra. Other fragments correspond to the loss of HCl in combination with Cl losses. Again, a characteristic and common fragment for the three unsaturated systems was a fragment at m/e 233 ($C_{13}H_7Cl_2$) reminiscent of the 235 fragment found for the saturated systems. DDE under photooxygenation conditions[17] yields compounds which have been characterized as dichlorofluorene derivatives resulting from photocyclization. It is also interesting that 4,4'-dichlorobiphenyl was detected among their products; however, a corresponding fragment (m/e 222) was not discernible in our normalized spectra. The dichlorofluorene systems are highly resonance stabilized and represent likely structures for the ions observed at m/e 233 and 235. A simple resonance stabilized dichlorobenzhydryl carbonium ion structure could also be drawn for the m/e 235 fragment which would not require a cyclic rearrangement. As would be expected, the parent ions of these unsaturated systems are

m/e 233   m/e 235   or

reasonably stable (DDE-57%, DDMU-38%, DDNU-83%) which facilitates their detection in complex mixtures.

The oxidative metabolite systems afford somewhat more unique and characteristic mass spectra since they are more varied in structure. The systems (DBH and DBP) containing oxygen functions at their benzhydryl carbon positions are characterized by the

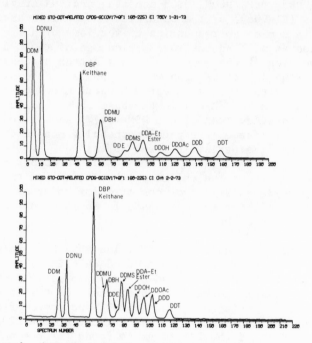

Fig. 2. RGC's (OV-17 + QF-1) of DDT compounds on interface to EI (top) and CI mass spectrometers.

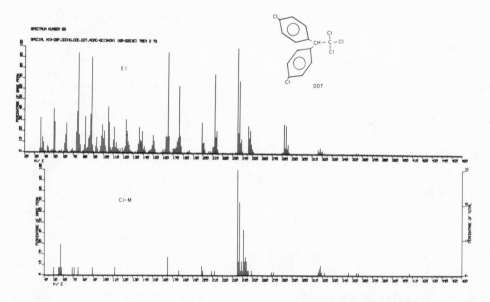

Fig. 3. Electron impact (70ev) and methane chemical ionization mass spectra of DDT.

p-chlorophenyl carbonyl (P-Cl$\emptyset \equiv$ O$^+$) ion at m/e 139 which occurs
as the base peak. The charge can likewise be carried to a lesser
extent by the p-chlorophenyl fragment (m/e 111). Other fragments
correspond to the loss of Cl, and for DBH, the loss of water as
well. The parent ion of the benzhydrol is weaker than that in DBP
(4% and 22%) since stabilization can be achieved by further fragmentation. It is interesting to note that DDOH shows some m/e 139
($\approx$7%) indicating a possible cyclic involvement of the benzhydryl
carbon and the hydroxyl group via an oxirane type ring system. In
both DDOH and the DDA-ethyl ester the m/e 235 fragment previously
discussed occurs again as the base peak, but in the acetate of DDOH
further fragmentation of this ion via Cl loss yields m/e 165 as the
base peak (m/e 235, $\approx$75%). In addition to the C$^\alpha$ - C$^\beta$ bond cleav-
ages, these latter two systems show the characteristic losses of
acetic acid and methylene acetate (for DDOH-acetate) and carbethoxy
(for DDA-Et ester) groups.

Figure 2B shows the RGC of the same mixture of DDT compounds
on the same column in the gas chromatograph interfaced to the
chemical ionization mass spectrometer system in which methane
functions as both carrier and reagent gas. The elution pattern for
these compounds observed with the EI system did not change; however,
somewhat shorter retention times were obtained with sharper peaks.
Such a chromatographic behavior would be expected since the CI
system has no separator and, therefore, there is less surface area
for adsorption to occur. One of the peculiarities of both EI and
CI mass spectrometry combined with GC, and possibly a disadvantage,
is the observation that depending on which point on the GC peak the
MS spectrum is taken, one can get appreciable variations in relative
intensities of fragments including changes in the base peak. Generally, the spectra on the ascending side of the peak are richer in
parent ions. Since all of the compounds studied were introduced in
essentially identical relative amounts ($\approx$1 µg), the spectrum chosen
for analysis in each case was that containing the largest total ion
percent for a given peak.

CI mass spectra are also characterized by the appearance of
recombination fragments involving the transfer of massive entities
such as protons, hydride ions, or alkylcarbonium (for methane, CH$_5^+$,
C$_2$H$_5^+$, and C$_3$H$_5^+$ are most important) ions. An (M-1)$^+$ quasi-
molecular ion is generally characteristic of hydrocarbons, but
depends to a large extent on the availability of hydride abstractable
hydrogen atoms and the stability of the resultant ions. For example,
many of the protons in the polycyclodiene pesticide systems to be
discussed are not suitable since bridge head carbonium ions would
result which are very unstable if not impossible to form. The
introduction of oxygen-containing functional groups, some of which
possess a permanent dipole, into these chlorinated hydrocarbon
systems can complicate the parent ion picture since such groups
tend to be proton acceptors. Therefore, the parent ion region may

be a combination of $(M-1)^+$ and $(M+1)^+$ ions as well as M^+ which
could make an exact assignment of molecular weight (when considering
parent ion region alone) difficult. No attempt has been made to
completely decipher the parent ion region except in cases where one
or the other process clearly has an advantage which would be characteristic of the system. An examination of the CI-MS spectra for
the saturated metabolite systems again reveals the presence of
fragments resulting from the loss of Cl and HCl as well as cleavage
of the $C^\alpha - C^\beta$ bond. But the most characteristic fragment found
for all of the saturated systems was that resulting from cleavage
of a p-chlorophenyl group which, in the case of DDT, could rearrange
to form a substituted tropylium ion. However, other reasonable
structures can be drawn such as p-quinoidal type systems which require no subsequent rearrangement of the initial cleavage product.
Such a fragmentation accounts for the base peaks of DDD, DDMS, DDM
and in some spectra for DDT as well. This apparently facile fragmentation process may reflect an electron rich carbon to phenyl bond
which is susceptible to attack by an alkyl carbonium ion. The

from DDT   or

fragment ion m/e 235 previously described was again present and intense in the spectra of all the systems except DDM where m/e 236
appears as the parent ion (≈8%). The parent ions in the other saturated systems were present in from 3 to 8 percent relative intensities. Figure 3 illustrates the differences in the EI and CI mass
spectra of DDT, a representative example of the saturated metabolite
systems.

   A comparison of the CI spectra of the unsaturated metabolite
systems again reveals the prevalence of the $(M-p-Cl\emptyset)^+$ fragment,
occurring as the base peak in DDMU and DDNU and of moderate intensity (≈16%) in DDE. In addition, it is somewhat surprising that
the fragment at m/e 233 which is characteristic of the EI spectra
of these systems is essentially absent from their CI spectra. The
parent ions are of moderate intensity for DDNU (15%) and DDMU (18%)
and some spectra of DDE showed $(M-1)^+$ as the base peak, although
the weaker spectra of DDE obtained from the GC run shown in Fig. 2B
contained a base peak of m/e 57. Figure 4 pictures the EI and CI
spectra of DDE for comparison.

   The oxidative metabolites again show somewhat more varied fragmentation patterns for their CI spectra reflecting specific

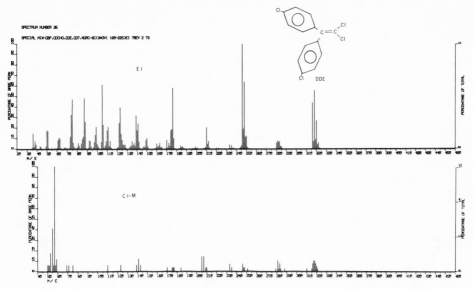

Fig. 4.   Electron impact (70ev) and methane chemical ionization mass spectra of DDE.

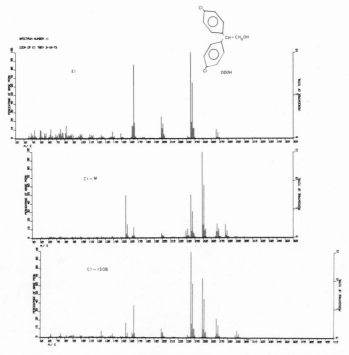

Fig. 5.   Electron impact (70ev) and methane and isobutane chemical ionization mass spectra of DDOH.

functionality. However, the base peaks for DBP, DBH, and DDA-ethyl
ester again correspond to the loss of the p-chlorophenyl group or
$(M-111)^+$. This fragment is also present in 50% and 4% of the relative intensities for DDOH and DDOH-acetate, respectively. DDOH
clearly shows an $(M+1)^+$ and DDOH-acetate and $(M-1)^+$ recombination
fragment. In addition, it was observed that both DDOH and its
acetate give rise to fragments which can only be explained by a
recombination-elimination process. For example, both DDOH and its
acetate show a fragment at m/e 277 which corresponds to a $C_2H_5^+$
recombination with elimination of water and acetic acid, respectively. Since recombination fragments are normally not appreciably
stable, the elimination process must generate a quite stable ion
(17% - DDOH and 11% for its acetate). In order to confirm the
suspected recombination-elimination process, the CI spectra of
DDOH and its acetate were obtained using isobutane as the reagent
gas. Since similar recombination processes for this gas involve the
addition of $C_3H_5^+$ ($\approx$40%) and $C_3H_7^+$ (100%), one would expect to see
new fragments at m/e 289 and 291 with the absence of the 277 fragment found for methane. Figure 5 shows the EI, CI-methane, and
CI-isobutane spectra of DDOH for comparison, and one indeed finds
that the recombination-elimination process does hold true for isobutane with the appearance of the suspected overlapping fragments
at m/e 289 and 291.

It is perhaps worthwhile to speculate on the structure of these
ions. It is reasonable to postulate the formation of substituted
and protonated cyclopropane ions, although the exact mechanism of
formation is unclear. The CI mass spectra for kelthane acetate show

m/e 277   m/e 289   m/e 291

that similar processes are operating in the chemical ionization of
this compound as well. The recombination-elimination processes
appear to be characteristic of those molecules in which an oxygen
containing leaving group is available that can be substituted on
either C^α or C^β of the ethane portion of the molecule. Protonated
cyclopropane has been previously proposed[18] in the ionization of an
isopropyl group. It is perhaps interesting to note that the insecticidal activity of several dichlorocyclopropanes substituted by
halogen and alkoxy groups in the phenyl rings follows fairly closely
the activity of the reported DDT analogues.[19]

As suspected, DDOH and its acetate show the characteristic respective losses of the hydroxyl and acetate groups including methylene acetate. These processes account for the base peaks of these two compounds with methane as reagent gas. The DDA-ethyl ester shows the expected loss of the carbethoxy group (RI, 31%).

It was also of interest to compare both the EI and CI mass spectra of p,p'-DDT with its o,p'-isomer as well as with the tri-fluoromethyl DDT analog. Figure 6 shows the EI spectra of o,p'-DDT and the trifluoromethyl analog of p,p'-DDT. The spectrum of o,p'-DDT closely resembles that of the p,p'-isomer, at least qualitatively. The normalized spectrum is absent of parent ion but contains the same base peak (m/e 235). Quantitative differences in the spectra would be expected since there is a somewhat greater tendency on the part of o,p'-DDT to lose a chlorine atom due to a greater degree of steric aggravation from the o-substituted chlorine atom. The F-DDT analog spectrum also shows the m/e 235 fragment as its base peak but shows an intense ($\approx$44%) parent ion at m/e 304. The CF$_3$ group is approaching a methyl group in steric size. The lesser steric requirements coupled with the increased stability of the C-F bonds probably accounts for the intense parent ion. The CI spectra of these systems as well are characterized by the fragment from loss of a chlorophenyl group. The drastically different polarizing effects of the trichloromethyl group as compared to the trifluoro-methyl group are evident in the RGC of the two compounds shown in Fig. 7.

CI mass spectrometry has been particularly helpful for char-acterizing reactive and unstable systems isolated from chemical reactivity investigations supporting diverse metabolism studies. For example, kelthane acetate (KOAc) affords EI spectra very similar to that found for DDE. Previous chemical studies[20] had indicated that KOAc decomposes under either thermal or photolytic influence to yield DDE as a major product. Figure 8 shows the RGC of KOAc along with DDE, DDT and DBP. However, CI spectra of KOAc show the expected fragments for the structure as drawn in Fig. 1. The base peak corresponds to the loss of acetic acid. Although the parent ion is not present (See Fig. 9), a recombination-elimination frag-ment at m/e 378 involving M-1$^+$, C$_2$H$_5^+$ and loss of acetic acid is present ($\approx$7%) in addition to other fragments which can only be ex-plained by invoking recombination-elimination fragmentation processes. As previously pointed out, such fragments from CI are clearly characteristic of the compounds containing good leaving groups at C$^\alpha$ or C$^\beta$.

A possible intermediary metabolite system of DDT investigated was DDCHO, the suspected aldehyde precursor of DDA. DDCHO undergoes slow decomposition with or without solvent to form DBP. However, MS spectra free of the ever present DBP contaminant can be obtained by GC-MS. The expected fragmentation pattern is found in the

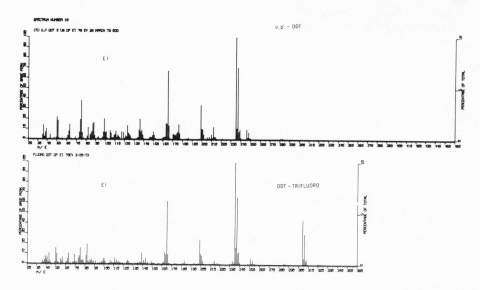

Fig. 6.   Electron impact (70ev) mass spectra of o,p'-DDT and the trifluoro analog of p,p'-DDT.

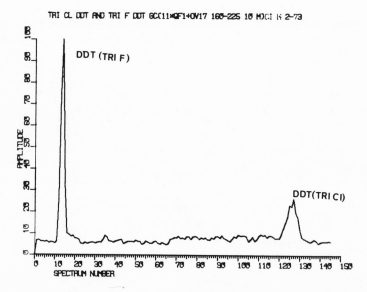

Fig. 7.   RGC (OV-17 + QF-1) of p,p'-DDT and its trifluoro analog.

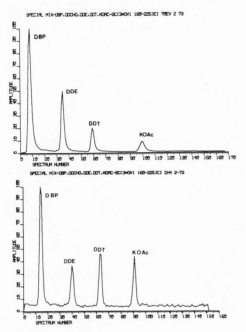

Fig. 8.   RGC (OV-1) of DBP, DDE, p,p'-DDT and kelthane acetate (KOAc).

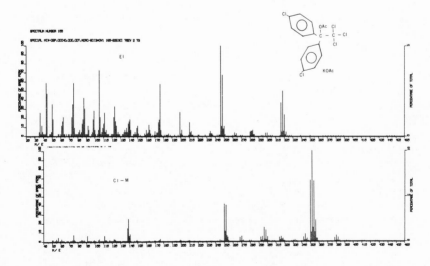

Fig. 9.   Electron impact (70ev) and methane chemical ionization mass spectra of KOAc.

EI-spectra of the aldehyde; however, CI-MS appears to be the most desirable approach for low level identification of the aldehyde since its CI-methane spectra show $(M+1)^+$ as the base peak. The aldehyde can be stabilized in the form of its enol ether acetate which affords an intense ($\approx$40%) $(M-1)^+$. However, the CI spectrum of the acetate was one of few which had a weak fragment resulting from p-chlorophenyl loss. This may possibly reflect poorer electron density about the carbon-phenyl bonds since electron flow is in the direction of the carbonyl group.

An extended version of the enol ether acetate of DDCHO was investigated (See Fig. 10) which provided an excellent example of the combined advantages of GC-MS in studying a very unstable system.[12] The compound rearranges in a cyclic fashion as pictured to produce a series of related products. Figure 11 shows the structures of these systems as deduced from their EI and CI mass spectra. It would be very difficult, if not impossible, to determine the presence of all these structures by any other means.

In a comparison of the CI quasi-molecular ion (with H^+ donating reagents) with the EI molecular ion region, CI generally offered no appreciable advantage or was at a disadvantage for the saturated and unsaturated type DDT metabolite systems. However, CI clearly had the advantage for the majority of the oxidative metabolite systems studied.

The discussion of the mass spectra of the chlorinated polycyclodiene systems will be divided into two groups. The first group includes aldrin, dieldrin, and isomers of dieldrin, viz., photodieldrin, aldrin ketone, endrin, and delta-ketoendrin. The second group includes the hydroxylic and ketonic metabolites of aldrin and dieldrin and some of their chemical derivatives. Figures 12-13 show the basic structures for most of the systems studied including some of the stereoselective synthetic methods[13] used in their preparation. The EI and CI spectra of these systems will be discussed together since they were qualitatively very similar. As was pointed out earlier, there seems to be no general rule regarding formation of a quasi-molecular ion in the CI spectra of these compounds and their derivatives. Both $(M-1)^+$ and to a lesser extent $(M+1)^+$ have been noted in their spectra. $(M+1)^+$ formation seems to be largely dependent on the presence of specific oxygen-containing functionality about the C4-C5 bond, e.g., the epoxide group. Detection of quasi-molecular ions for these systems containing six Cl atoms is further complicated by the isotopic distribution pattern.

Very few stable ions are formed in the initial stages of fragmentation of most of these systems and, therefore, quite extensive fragmentation usually results. The lower energy requirements of chemical ionization mass spectrometry favor less extensive

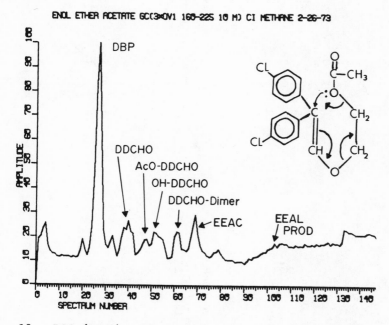

Fig. 10.   RGC (OV-1) of 1,1-<u>Bis</u>(<u>p</u>-chlorophenyl)-3-oxa-1-pentenyl-
5-acetate (EEAC) and its decomposition products.

Fig. 11. Compounds derived from enol ether acetate (EEAC) rearrangement.

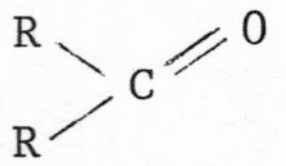

DBP
$C_{13}H_8OCl_2$
250

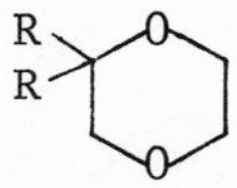

DDCHO
$C_{14}H_{10}OCl_2$
264

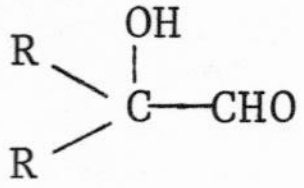

OH–DDCHO
$C_{14}H_{10}O_2Cl_2$
280

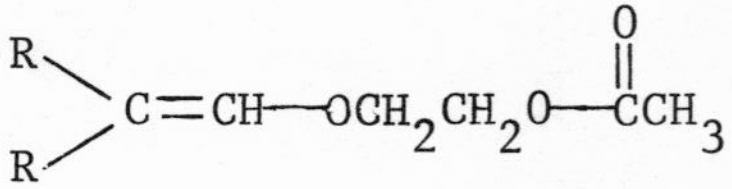

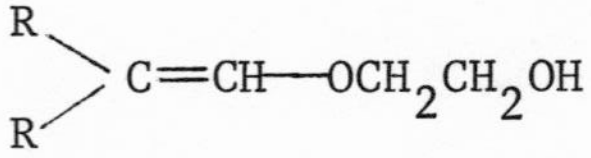

EEAL
$C_{16}H_{14}O_2Cl_2$
308

EEAL PROD
$C_{16}H_{14}O_2Cl_2$
308

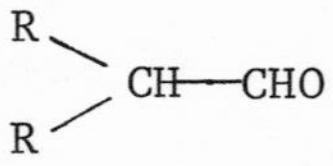

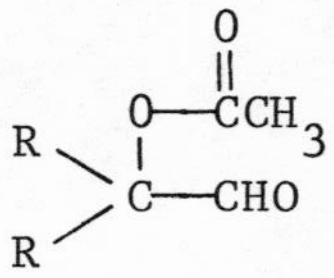

AcO–DDCHO
$C_{16}H_{12}O_3Cl_2$
322

EEAC
$C_{18}H_{16}O_3Cl_2$
350

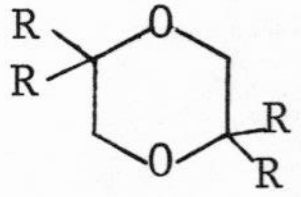

DDCHO–DIMER
$C_{28}H_{20}O_2Cl_4$
528

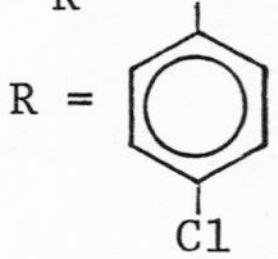

Abbrev. Title,
Formula, Molecular wt.
based on Cl^{35}

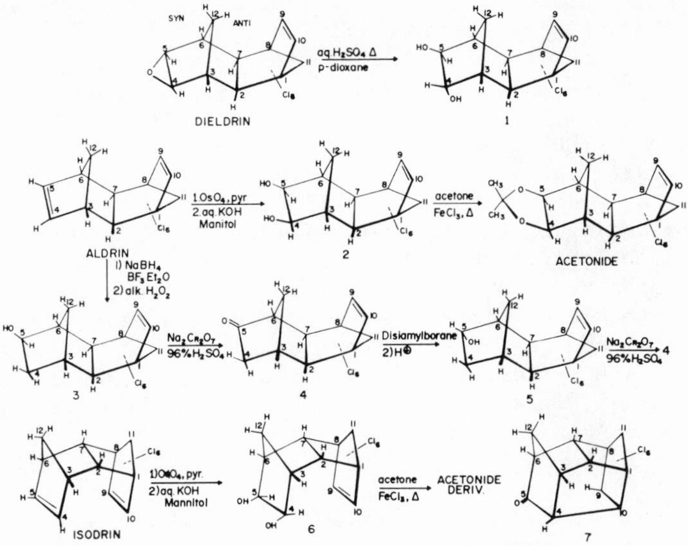

Fig. 12. Structures and synthetic procedures for some chlorinated polycyclodiene pesticide metabolites.

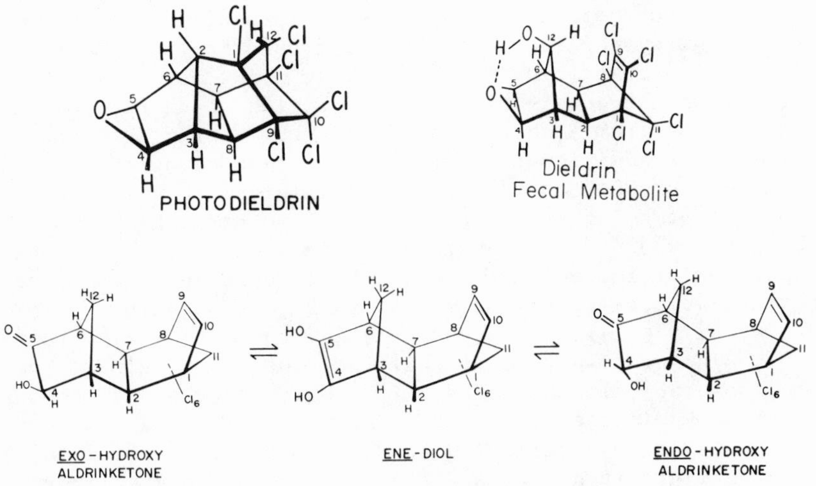

Fig. 13. Structures of special interest polycyclodiene systems.

fragmentation and a predominance of the initial characteristic
fragmentation processes. The salient features of the spectra are
ions corresponding to a set of retro-Diels Alder processes, ions
resulting from successive losses of Cl, HCl, or both, ions produced
by combinations of a retro-Diels Alder process with a preliminary
or subsequent loss of Cl or HCl, and ions more specifically involving
the various groups substituted at the C4-C5 positions. The EI mass
spectra of some of the parent polycyclodiene pesticides have been
previously investigated.[21]

An understanding of the two sets of retro-Diels Alder processes
can be acquired using aldrin as the basic system for study. Depending on which bonds are broken at the ring fusion cyclopentadiene
(m/e 66) and hexachlorocyclopentadiene (does not appear to ever
carry the charge) can be formed which are accompanied by one or the
other norbornadiene system, hexachloronorbornadiene (m/e 296) and
norbornadiene itself (m/e 92). It is likely that these latter
systems rearrange[22] to form tropylium ions. Other ions seem to be
the result of a retro-Diels Alder process in combination with a
chlorine transfer process. The base peak in the EI mass spectrum
of aldrin is found to be m/e 66. However, the CI mass spectrum of
aldrin, which shows a quasi-molecular ion $(M-1)^+$, has a base peak
of m/e 326 corresponding to the loss of HCl and indicates less
extensive fragmentation.

Dieldrin, in comparison, yields an ion at m/e 79 (100%) in both
its EI and CI spectra, which occurs as the base peak in several
other systems related to dieldrin. Apparently, the favored retro-
Diels Alder process is the one which affords the norbornadieneoxide
which undergoes further fragmentation with loss of CHO to produce
a cyclohexadiene system $(C_6H_7^+)$. Oxodihydroaldrin (aldrin ketone)
yields EI spectra containing m/e 79 as base peak, but in its CI-
methane spectrum the quasi-molecular ion region contained the base
peak. The CI spectrum also contained the characteristic recombination fragments in addition to a fragment corresponding to the
combined losses of HCl and CH_2CO. Photodieldrin, on the other hand,
affords a base peak ion at m/e 81 in its EI spectra which can
reasonably be explained by a protonated cyclopentadienone structure.
The CI spectrum of photodieldrin did not yield an appreciably more
intense quasi-molecular ion but contained a base peak in the $(M-Cl)^+$
pattern (See Figure 14). A characteristic fragment is also found
at m/e 163. The exact nature of this fragment is not clear; however,
it resembles a two chlorine pattern ion, which suggests a protonated
dichlorocyclohexadienone structure. One would expect a somewhat
reduced tendency of photodieldrin to undergo the retro-Diels Alder
decomposition since it has an additional bridging bond to break.
Dieldrin also contains the m/e 81 ion in its CI spectra suggesting
that the quasi-molecular ion $(M-1)^+$ formed may resemble a photodieldrin type structure (See Figure 15).

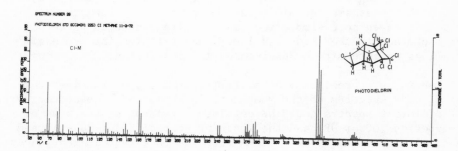

Fig. 14.   Methane chemical ionization mass spectrum of photo-dieldrin.

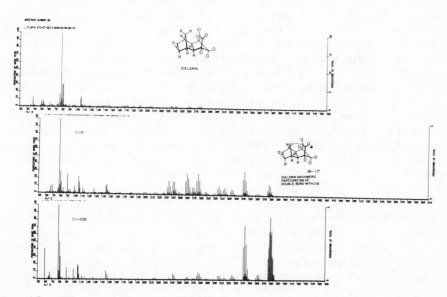

Fig. 15.   Electron impact (70ev) and methane and isobutane chemical ionization mass spectra of dieldrin.

Almost all of the polycyclodiene systems examined showed appreciably intense ions in the M—Cl region. It is interesting to speculate that one of the dichloroethylene Cl atoms is lost first since photolysis of aldrin and dieldrin have yielded such des-chloro compounds as products.[23] Figure 15 shows EI and CI spectra of dieldrin as a representative example of the compounds discussed so far. Both CI-methane and CI-isobutane spectra are shown in which one sees an even more simplified dieldrin fragmentation pattern when isobutane is used (an even lower energy ionization process) as well as an intensified quasi-molecular ion region.

Endrin, a stereoisomer of dieldrin, by far gave the most complex EI mass spectrum of the compounds investigated. Quantitatively very different spectra could be obtained depending on the mode of introduction (GC or DP) and the ionizing energy (20 or 70 ev) used. The complexity of the system stems from the fact that it rearranges readily[24] under a variety of conditions to form at least two other products, delta-ketoendrin (base peak, EI, m/e 67; and CI, m/e 343) and an endrin aldehyde. The mass spectrum of endrin appears to contain fragments indicative of at least three structures of which delta-keto endrin (DKE) seems to be a major contributor. Utilizing the computer to subtract DKE spectra away from endrin spectra, a fragment at m/e 261 ($C_7H_2Cl_5$) stands out and appears to be a characteristic fragment of the true endrin structure. In a similar fashion certain fragments can be shown to be characteristics of DKE such as the fragment at m/e 315 (loss of CO,Cl). Again, utilizing the computer, and the 315 fragment, one can determine the percentage of total ions representing DKE in an endrin spectrum under the conditions used. This, in turn, is a measure of the rearrangement propensity of endrin to DKE under a given set of conditions. Figure 16 again illustrates the advantages of chemical ionization for producing spectra of endrin in which rearrangement processes are minimized and, at the same time, that are rich in quasi-molecular ion $(M+1)^+$.

Figure 17 shows the EI and CI-methane mass spectra of an actual biological metabolite sample obtained from dieldrin metabolism in the rat, viz., the major fecal metabolite (C-12 hydroxydieldrin). The EI spectrum is absent of parent ion, but the CI spectrum clearly indicates a quasi-molecular ion $(M+1)^+$ indicative of a hydroxylated dieldrin molecule. An apparent thermal and photolytic instability of this system has been previously noted[15] and would possibly account for the difficulty in seeing a parent ion with EI.

The second group of systems of the oxidative type include the aldrin alcohols (exo and endo-isomers), the aldrin diols cis and trans-isomers), and two oxidation products of the cis-aldrindiol. These systems were generally analyzed as their silyl ether derivatives since better chromatographic properties resulted. The derivatives were chromatographed as mixtures along with aldrin,

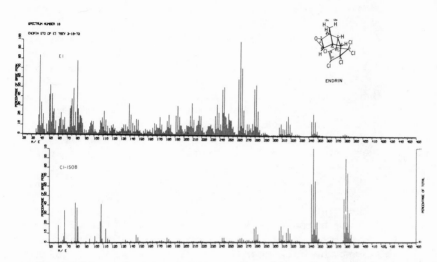

Fig. 16. Electron impact (70ev) and isobutane chemical ionization mass spectra of endrin.

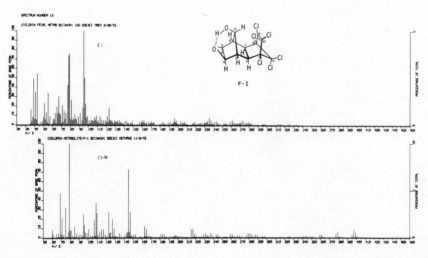

Fig. 17. Electron impact (70ev) and methane chemical ionization mass spectra of the major fecal metabolite (in rat) of dieldrin.

dieldrin, and the aldrin ketone on two different columns since
complete separation could not be achieved on one column. Figure 18
shows the RGCs of this mixture on the gas chromatograph interfaced
to the electron impact mass spectrometer. Similar RGCs were obtained on interface to the CI instrument. The normalized EI mass
spectra of the silyl ether derivatives of the aldrin alcohols are
not very informative since the base peak at m/e 73 corresponding
to the trimethyl silyl group accounts for 16-17% of the total ion
yield. However, with the aid of computer limited mass searches and
amplitude expanded spectra, one can discern the usual fragment ions
resulting from loss of Cl, HCl, and retro-Diels Alder processes.
In the CI spectra of the alcohols, the base peak at m/e 73 accounts
for 3-4% of the total ion yield, and the quasi-molecular ions
(m/e 452) are very evident (8-9%). In addition, ions more specifically involving the trimethyl silyl group are now evident, viz.,
$(M-CH_3)^+$ and (M-HOTMS and $HCl)^+$.

     Likewise, the trans and cis-aldrindiol disilyl ethers afford
rather indescriptive normalized EI spectra. Again, the CI spectra
are much richer in ions at or near the quasi-molecular ion region.
The $(M-Cl)^+$ becomes the base peak for both diols and both quasi-
molecular ions are evident (trans, 2% and cis, 10%). If one
considers the M-Cl fragment for the diols and assumes that the
disilyl ethers have a similar tendency to lose chlorine, then the
total ion yields for this fragment should be a reflection of the
stabilities of the two compounds. The M-Cl total ion yields for
the cis and trans diols were 6.5 and 10.2, respectively. This would
imply a greater stability for the trans-disilyl ether which would
be expected on the basis of steric arguments. Although mass
spectrometry is normally not desirable for the assessment of stereochemistry,[25] it has some value when applied to rigid polycyclic
systems such as these and differences are expressed in terms of
total ion yields. Figure 19 shows the expanded EI mass spectrum
and the CI-methane spectrum of the trans-diol disilyl ether for
comparison.

     Chemical ionization mass spectrometry has also been helpful
in examining complex product mixtures from organic chemical reactivity studies involving various polycyclodiene systems. For
example, oxidation studies of the cis-aldrindiol afforded two
related and apparently isomeric products, one of which behaved like
the suspected hydroxyketone. The mixture was analyzed by GC-MS
using the CI system, and the spectra obtained are shown in Fig. 20
along with the RGC. The mass spectrum of the major component
showed a quasi-molecular ion $(M-1)^+$ at m/e 393 (22%) and a base
peak of m/e 95 (C_6H_7O). In addition, ions were present corresponding to the loss of CHO, CHOH and HCl, and CHO-CHOH-Cl. On the
basis of these fragments, the hydroxy ketone structure was assigned.
The other component had a quasi-molecular ion $(M+1)^+$ at m/e 395 and
a base peak of m/e 79 (C_6H_7). Other fragment ions corresponded to

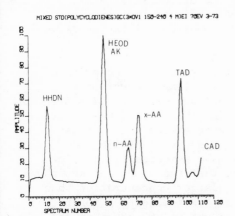

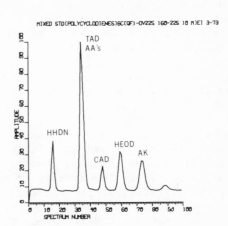

Fig. 18. RGC's (OV-1) of aldrin, dieldrin, and aldrin ketone mixed with the silyl ether derivatives of the aldrin alcohols and diols.

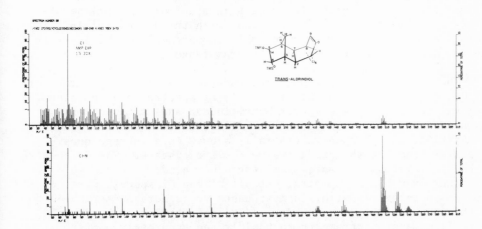

Fig. 19. Electron impact (70ev, 1% expanded by 20) and methane chemical ionization mass spectra of the silyl ether derivative of trans-aldrindiol.

losses of CHO and 2Cl, and $C_2H_2O_2$ and 2Cl. The spectrum was both qualitatively and quantitatively different from that of the other component, and, therefore, this compound was assigned the ene-diol structure. Preliminary work on the GC-MS of the silyl ether derivatives of these two compounds has confirmed that one isomer incorporates at least two silyl ether groups on silylation as would be expected for the ene-diol.

Chemical reaction studies involving opening of the oxirane ring of dieldrin have afforded some interesting product mixtures. Figure 21 shows the RGC of the GC-MS analysis with the CI system for one of these product mixtures. Utilizing the limited mass search capability of the computer, one is able to assign two of these peaks to the trans- and cis-aldrindiol diacetates at m/e 480. Authentic samples of these diacetates gave essentially identical GC-MS data. It is interesting to note that the total ion RGC shows that the cis-diacetate is present in a larger concentration than the trans; however, the LMS-RGC for the quasi-molecular ion region indicates that about 2-1/2 times the amount of 480-486 ions are derived from the trans-diacetate. Here again is good evidence that the cis-oriented groups are more sterically compressed and, therefore, comprise a more unstable arrangement. In addition, the total ion yield for the M-HOAc fragment is correspondingly higher for the cis-diacetate as would be expected.

Although sensitivity (minimum detectability) was not a major concern in these studies, the majority of the mass spectra obtained for all of the compounds discussed was on a sample size of about one microgram and with the instrument in a normal (medium) sensitivity range setting. It is felt that with the aid of multiple ion detection (MID) the minimum detectable amounts will be at least in the low nanogram or even in the picogram range.

The CI spectra of the DDT compounds were both qualitatively and quantitatively different from their EI spectra and, therefore, chemical ionization mass spectrometry indeed adds a new dimension to the MS identification and characterization of these compounds. In addition, the CI spectra generally showed more intense quasi-molecular ions for the oxidative metabolite systems. The CI spectra of the polycyclodiene systems, on the other hand, were not so different from the EI spectra, but all of the CI spectra were simpler to interpret with, in most cases, considerable enrichment of the important high mass end. Additional studies are underway to utilize mass spectrometry to rapidly and accurately confirm the presence of the C-12 dichloroethylene bridge characteristic of photodieldrin and its ketonic metabolite. In addition, mass spectrometry should be of further value in assessing stereochemistry of these rigid systems where the quantities of compounds available will not allow more conventional methods to be employed. The EI mass spectra of several of these systems contained fragments indicative of their thermally and photolytically derived products. Therefore, the EI mass spectrometer continues to be a reliable model system for measuring such unimolecular processes.

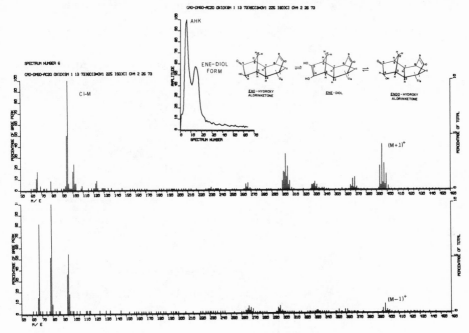

Fig. 20.   CI-methane mass spectra of <u>cis</u>-aldrindiol oxidation products.

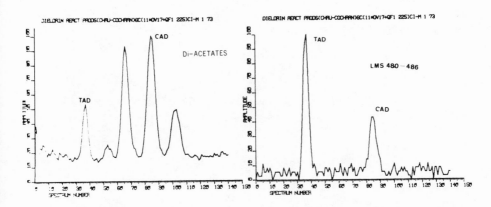

Fig. 21.   Total ion RGC and limited mass RGC of dieldrin acid catalyzed acetylation products.

References

1. R. O. Mumma and T. R. Kaniner, J. Econ. Entomol., **59**(2), 491 (1966).
2. O. Hutzinger and W. D. Jamieson, Bull. Environ. Contam. Toxicol., **5**(6), 587 (1970).
3. O. Hutzinger, W. D. Jamieson and V. Zitko, Nature, **226**(5246), 664 (1970).
4. U. Solang, C. M. Himel and T. Dirks, Bull. Environ. Contam. Toxicol., **8**(2), 97 (1972).
5. O. Hutzinger and S. Safe, in "Mass Spectrometry of Pesticides", G. Zweig, Editor, CRC Press, Cleveland, Ohio, 1973, pp. 225.
6. F. J. Biros, R. C. Dougherty and J. Dalton, Org. Mass Spectrum, **6**(11), 1161 (1972).
7. M. S. B. Munson and F. H. Field, J. Am. Chem. Soc., **88**, 2621 (1966).
8. P. C. Rankin, J. Ass. Offic. Anal. Chem., **54**(6), 1340 (1971).
9. R. C. Dougherty, J. Dalton and F. J. Biros, Org. Mass Spectrum, **6**(11), 1171 (1972).
10. H. M. Rosenstock and M. Krauss, in "Mass Spectrometry of Organic Ions", F. W. McLafferty, Editor, Academic Press, New York, N. Y., 1963, p. 2.
11. J. E. Peterson and W. H. Robison, Toxicol. Appl. Pharmacol., **6**, 321 (1964).
12. J. D. McKinney, R. E. Hawk, E. L. Boozer and J. E. Suggs, Can. J. Chem., **49**(23), 3877 (1971).
13. J. D. McKinney, L. H. Keith, A. Alford and C. E. Fletcher, Can. J. Chem., **49**(12), 1993 (1972).
14. J. D. Albright and L. Goldman, J. Am. Chem. Soc., **89**(10), 2416 (1967).
15. J. D. McKinney, H. B. Matthews and L. Fishbein, J. Agric. Food Chem., **20**(3), 597 (1972).
16. A. S. Y. Chau and W. P. Cochrane, Bull. Environ. Contam. Toxicol., **5**(6), 515 (1971).
17. J. R. Plimmer, U. I. Klingebiel and B. E. Hummer, Science, **167**(3914), 67 (1970).
18. P. N. Rylander and S. Meyerson, J. Am. Chem. Soc., **78**, 5799 (1956).
19. G. Holan, Nature, **221**, 1025 (1969).
20. J. D. McKinney and L. Fishbein, Chemosphere, **1**(2), 67 (1972).
21. J. N. Damico, R. P. Barron and J. M. Ruth, Org. Mass Spectrum, **1**, 331 (1968).
22. P. Story, J. Org. Chem., **26**(2), 287 (1961).
23. G. L. Henderson and D. G. Crosby, J. Agric. Food Chem., **15**, 888 (1967).
24. W. L. Burton, Ph.D. Dissertation, No. 72-30, 864, 1972, pp. 106.
25. K. Biemann and J. Seibl, J. Am. Chem. Soc., **81**, 3149 (1959).

POSITIVE AND NEGATIVE CHEMICAL IONIZATION MASS SPECTRA OF POLY-

CHLORINATED PESTICIDES[1]

Ralph C. Dougherty, J. David Roberts, Harvey P. Tannenbaum

Department of Chemistry, Florida State University,
Tallahassee, Florida   31306

Francis J. Biros

Primate and Pesticides Effects Laboratory, Environmental
Protection Agency, Perrine, Florida   33157

INTRODUCTION

Chemical ionization (CI) and negative chemical ionization (NCI)
mass spectra generally show substantially less fragmentation than
their electron impact (EI) counterparts.[2] Consequently, the relative
intensity of ions with high information content, e.g. MH^+, is often
higher in CI than in EI spectra. The simplicity and sensitivity of
the CI spectra of drugs have resulted in the successful use of chemical ionization in the screening of crude extracts of biological
fluids for drugs and poisons.[3] The extension of these methods to
pesticide residue analysis is obvious but not necessarily direct.

If suitable procedures for CI and NCI examination of pesticide
residues in extracts of environmental substrates including human
tissues can be developed it should be possible to substantially reduce
the time required for an individual residue analysis. CI and NCI
spectra can be obtained and interpreted for a given sample in a matter
of minutes as compared to the time required to prepurify extracts and
subject them to gas chromatographic (glc) analysis. CI and NCI
analysis offer the additional advantage of positive molecular identification of specific residues. The early confusion between the PCB's
and DDT type compounds which was due to similar glc retention times,
would not have occurred if the analytical procedure had included
mass spectrometry.

The application of CI and NCI mass spectrometry to the analysis of pesticide residues in crude extracts of environmental substrates is not as straightforward as the application of these techniques to drug screening. The reason for this stems from the fact that the molecular sensitivity of different classes of pesticides to CI and NCI mass spectrometry can vary greatly. By comparison, the sensitivity of the various classes of drugs to CI is relatively constant. One example of structure dependent sensitivity of pesticides to CI mass spectrometry is the difference observed between the aromatic (DDT type) and polycyclic insecticides. DDT and its analogues give intense CI spectra with both isobutane or methane as the reactant gas. In contrast, many of the polycyclic insecticides do not respond to isobutane chemical ionization, presumably because their proton affinity is less than that of isobutylene. In CI spectra, the variation in sensitivity due to variations in proton affinity can be minimized by using methane as the reactant gas. CH_5^+ will exothermically transfer a proton to virtually any molecule.

NCI mass spectra are also subject to structure dependent variations in sensitivity. The three major processes for forming primary negative ions in NCI spectra are (1) resonance capture, (2) dissociative resonance capture and (3) ion pair formation.[5]

resonance
capture $\qquad M + e_s \longrightarrow M^{-\cdot *} \longrightarrow M^{-\cdot}$ (1)

dissociative
resonance capture $\qquad M + e_s \longrightarrow A^- + B$ (2)

ion pair
formation $\qquad M + e \longrightarrow M^* + e \longrightarrow A^- + B^+$ (3)

The primary ions may remain in the source long enough for analysis or they may subsequently react to give secondary ions, in a manner exactly analogous to CI mass spectra. Assuming that the distribution of secondary electron energies in the source does not vary with the sample, the relative susceptibility of a compound to processes (1), (2), or (3), will be highly structure dependent. The same is true of the susceptibility of a compound to subsequent ion molecule reactions. For example, the methylene chloride NCI mass spectra of commercial aromatic aldehydes are generally dominated by ions which correspond to $ArCOOH \cdots Cl^-$. The chloride ions are formed by dissociative resonance capture with methylene chloride. The heat of attachment of the aldehydes to chloride is so small that the addition reactions do not generally appear at normal operating temperatures. The heat of attachment of the carboxylic acids to chloride is, however, high, so low level acid impurities dominate the NCI spectra.

The variation of pesticide CI or NCI sensitivity with structure is a definite disadvantage for routine screening; however, the
disadvantage is small when compared to the simplicity and speed of
the CI and NCI analytical techniques. Furthermore, if both CI and
NCI spectra are used it is likely that a given compound will have a
high molecular sensitivity for at least one of the techniques.

In the following paragraphs we will review the CI and NCI mass
spectra of polycyclic[7,8] and DDT type[4] pesticides as well as present some preliminary results for the polychlorinated biphenyl compounds (PCB). Finally, we will illustrate the application of CI and
NCI mass spectrometry to screening partially purified human tissue
extracts for PCB residues.

POLYCYCLIC CHLORINATED PESTICIDES

Polycyclic chlorinated pesticides generally do not give intense
CI mass spectra with isobutane as the reagent gas because the proton
affinity of the pesticides is somewhat lower than that of isobutylene.
Therefore, proton transfer reaction (4) will not proceed.

$$M \quad + \quad \underset{\underset{CH_3}{\diagdown}\overset{\diagup}{CH_3}}{\overset{\overset{CH_3}{|}}{C+}} \quad \xrightarrow{\quad\times\quad} \quad MH^+ \quad + \quad \underset{\underset{CH_3}{\diagdown}\overset{\diagup}{CH_3}}{\overset{\overset{CH_2}{||}}{C}} \qquad (4)$$

Since the t-butyl cation is the dominant ion (>90%) in the high pres-
mass spectrum of isobutane a CI system of stronger acids had to be
employed. Methane is the obvious choice. The dominant ions in the
high pressure mass spectrum of methane are CH_5^+ and $C_2H_5^+$; both of
these ions exothermically protonate the polycyclic pesticides. protonation of a double bond or oxygen in the polycyclic system will
generally result in the formation of a stable MH^+ ion or favor ring
fragmentation by a retro-Diels Alder reaction or similar process.
Protonation of chlorine will generally result in elimination of HCl
to give $(M-Cl)^+$ ions. The CI mass spectrum of aldrin (Fig. 1) and
nonachlor (Fig. 2) are representative of the polycyclic pesticides.
These spectra suggest that Field's model of "random attack-local-
ized reaction" which has been successfully applied to the methane
CI spectra of saturated hydrocarbons[2], is applicable in this instance. The dominant importance of $(M-Cl)^+$ ions in the spectra of
this class of compounds is consistent with random protonation on
chlorine followed by elimination of HCl.

The CI spectra of both aldrin and nonachlor show fragment ions
that can be attributed to retro-Diels Alder reactions of the $(M-Cl)^+$
ion. In the aldrin spectrum this ion appears at m/e 261, and is

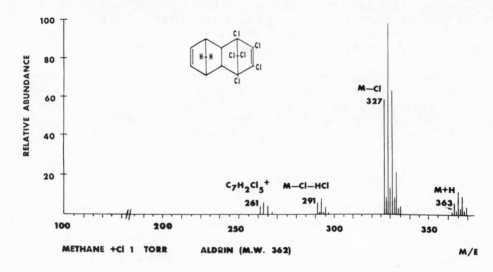

Figure 1.   Positive methane chemical ionization mass spectrum
             of the pesticide aldrin. (*Organic Mass Spectrometry*)

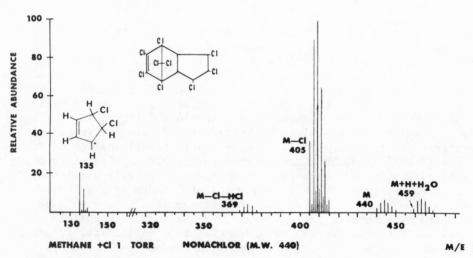

Figure 2.   Positive methane chemical ionization mass spectrum
             of nonachlor. (*Organic Mass Spectrometry*)

presumably the internal π-complex, I.

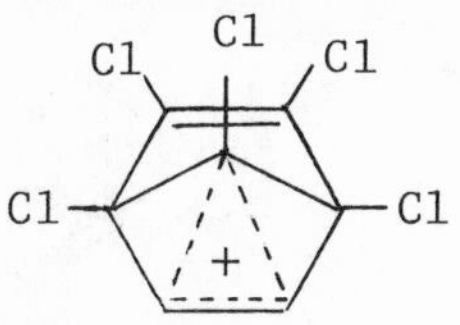

I

It is interesting that in the retro-Diels Alder decomposition of both aldrin and nonachlor the charge remains almost exclusively with the ene-half of the Diels Alder pair. This is almost certainly the result of the fact that ions such as I are delocalized aromatic systems while the cyclopentadienyl cation would be antiaromatic.

Introduction of an oxygen atom as in heptachlor epoxide, II, increases the proton affinity of the system, and consequently increases the importance of MH^+ ions in the spectra. Nevertheless, a dominant feature in methane CI spectra of all of the polycyclic pesticidal compounds we have examined thus far is the $(M-Cl)^+$ ion.

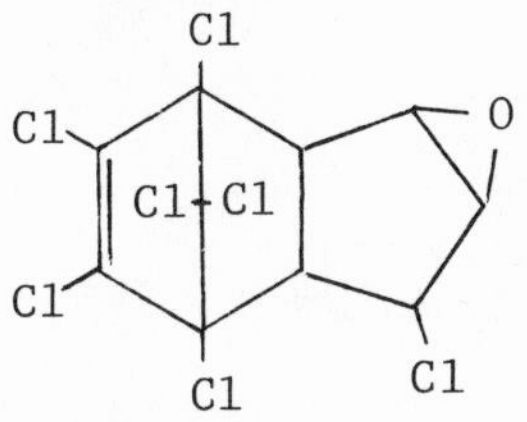

II

The methane NCI spectra of aldrin and nonachlor are shown in Figs. 3 and 4, respectively . The NCI spectrum of aldrin is one of the most complicated that we have observed to date. This is in large measure due to absorbed water which gives rise to oxygen containing ions in the NCI spectrum. A flow sheet for the observed NCI reactions of aldrin is shown in Fig. 5. Ion molecule attachment reactions are shown to the left while resonance capture, dissociative resonance capture and ion molecule displacement reactions are shown to the right in the figure.

The significant feature of the NCI spectra of polycyclic pesticides is the prominence of $(M+Cl)^-$ in every spectrum[8] . Thus, a suspected residue of aldrin which gave an ion cluster at m/e 327 in the CI spectrum should also give an ion cluster at m/e 397 in the NCI spectrum.

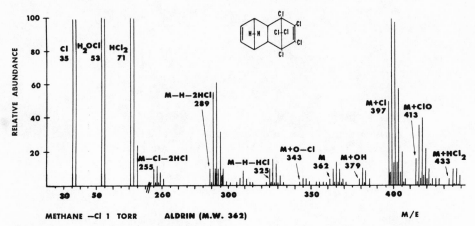

Figure 3.   Negative methane chemical ionization mass spectrum
of the pesticide aldrin. (*Organic Mass Spectrometry*)

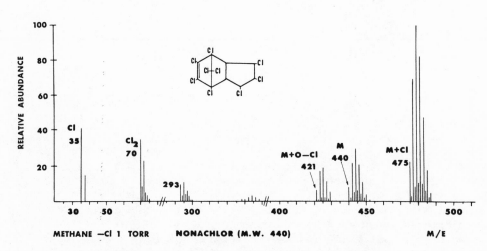

Figure 4.   Negative methane chemical ionization mass spectrum
of nonachlor. (*Organic Mass Spectrometry*)

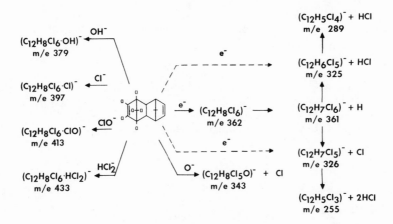

Figure 5.   Negative methane chemical ionization mass spectral
processes observed for the pesticide aldrin.
(*Organic Mass Spectrometry*)

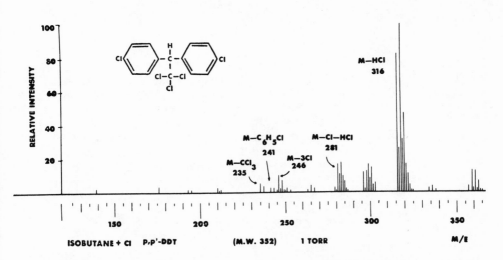

Figure 6.   Positive isobutane chemical ionization mass spectrum
of p,p'-DDT.

CHLORINATED AROMATIC PESTICIDES

The aromatic pesticides and metabolites in the DDT series have a large enough proton affinity to give intense CI spectra with isobutane as the reagent gas. Commercial isobutane is considerably cleaner and dryer than commercial methane, and as a result, isobutane NCI mass spectra show fewer oxygen containing ions than methane NCI spectra. For these reasons we used isobutane as the reagent gas in our initial studies of aromatic pesticidal compounds[4]. For a general survey application methane will have to be the reagent gas; preliminary experiments suggest that this change in reagent gas will not make a qualitative difference in the observed CI spectra or NCI spectra.

The isobutane CI spectra of p,p'-DDT, DDMU and p,p'-DDE are presented in Figs. 6, 7 and 8, respectively. The base peak in the p,p'-DDT CI spectrum (Fig. 6) corresponds to the p,p'-DDE molecule ion. The next most intense ion is observed at m/e 317, i.e. $(M-Cl)^+$. These same peaks dominate the p,p'-DDE spectrum (Fig. 8). The direct correspondence of the ions in these spectra would be an analytical disadvantage if it weren't for the results observed for the NCI spectra (see below).

DDMU, in addition to the other diphenylethylene compounds in this series exhibited an intense ion cluster corresponding to both $M^{+\cdot}$ and MH^+. Adducts such as $(M+C_3H_7)^+$ were also more prominent in the CI spectra of the ethylenic compounds than in those of their saturated analogues.

The isobutane NCI spectra of p,p'-DDT, DDMU and p,p'-DDE are illustrated in figs. 9, 10 and 11, respectively. The NCI spectrum of DDMU is exceptional in that $(M+Cl)^-$ is not the base peak, rather, $(M+HCl_2)^-$ is. This means that the major DDMU ion in the NCI spectrum will directly overlap with the $(M+Cl)^-$ ion from the DDD isomers . The positive CI spectra of DDMU and the DDD compounds are, however, distinct so both could be identified if CI and NCI spectra were used for screening extracts for pesticide residues. The confusion between DDT and DDE in the CI spectra of these compounds is virtually eliminated in the NCI spectra. The base peak in both cases is $(M+Cl)^-$, at m/e 387 and m/e 351, respectively.

POLYCHLORINATED BIPHENYLS

The PCB compounds constitute a class of chlorinated materials that are ubiquitous in the environment and may be found in substantial concentrations in human and animal (fish and wildlife) tissues as well as other environmental media. Under chemical ionization conditions there is little that a PCB molecule can do except add a proton to give MH^+ ions. Fig. 12 illustrates the methane CI spectrum of

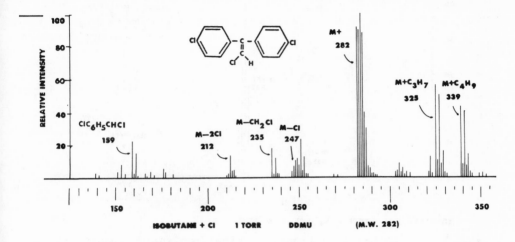

Figure 7.   Positive isobutane chemical ionization mass spectrum
of DDMU.

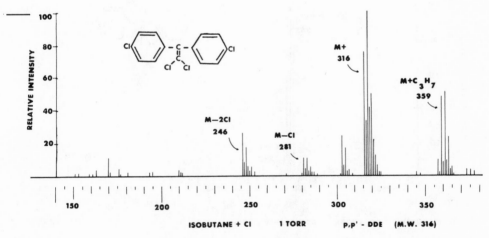

Figure 8.   Positive isobutane chemical ionization mass spectrum
of p,p'-DDE.

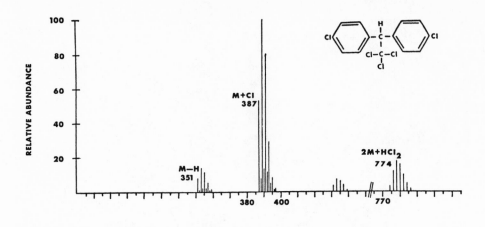

Figure 9.   Negative isobutane chemical ionization mass spectrum
             of p,p'-DDT.

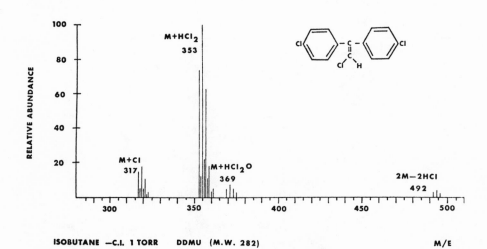

Figure 10.   Negative isobutane chemical ionization mass spectrum
             of DDMU.

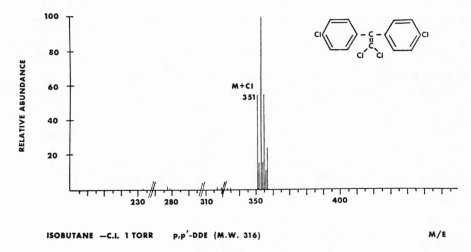

Figure 11.   Negative isobutane chemical ionization mass spectrum
             of p,p'-DDE.

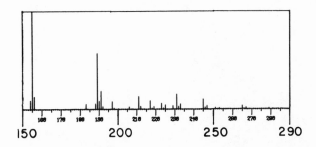

Figure 12.   Positive methane chemical ionization mass spectrum
             of Aroclor 1232 (mixture of PCB compounds with aver-
             age substitution of 1.9 chlorine atoms per molecule).

a commercial PCB formulation, Aroclor 1232, with an average substitution of 1.9 chlorine atoms per molecule. The spectrum shows MH^+ ions for biphenyl, m/e 155, chlorobiphenyl, m/e 189, and dichlorobiphenyl, m/e 223, in addition to hydrocarbon ions and ions containing one chlorine atom. The NCI spectrum of the same Aroclor taken under identical conditions of temperature and pressure is shown in Fig. 13. The ion at m/e 222 corresponds to the molecular anion of dichlorobiphenyl. Chloride attachment ions are found at m/e 257, m/e 291 and m/e 325 for the di, tri, and tetrachlorobiphenyls. The ion series m/e 237, m/e 271, m/e 305, m/e 339 and m/e 373 corresponds to $(M-19)^-$ for the tri, tetra, penta, hexa and heptachlorobiphenyls, respectively. Independent evidence suggests that the $(M-19)^-$ anion is formed by reaction of the molecular anion with oxygen (O_2) to produce the observed ion and a $ClO \cdot$ radical[8]. No indication of biphenyl or monochlorobiphenyl was found in this NCI spectrum. The molecular sensitivity for the process is obviously a function of the extent of halogenation of the biphenyl molecules. Table 1 illustrates a comparison of intensities of protonated molecular ions in the CI spectrum and $(M-19)^-$ ions in the NCI spectrum of Aroclor 1262, a commercial formulation of PCB containing an average substitution of 6.8 chlorine atoms per biphenyl molecule.

Table 1.

Aroclor 1262-Methane CI
Relative Intensities of $^{35}Cl_n$ Ions (First Isotope Peak)

Cl content in M	MH^+ (+8kV)	$(M-19)^-$ (-8kV)
Cl_9	–	0.6
Cl_8	2.8	11.
Cl_7	18.	51.
Cl_6	52.	44.
Cl_5	30.	3.6
Cl_4	16.	–
Cl_3	64.	–
Cl_2	60.	–
Cl	21.	–
Cl_0	10.	–

The precise origin of the $(M-19)^-$ anions and the molecular sensitivity of individual PCB compounds to CI and NCI examination are currently under investigation.

PCB 5700B (CL 1.9)   METHANE -CI:0.5 TORR,105°

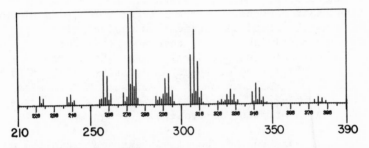

Figure 13. Negative methane chemical ionization mass spectrum
of standard Aroclor 1232.

TISSUE EXTRACT 1101 (1.2 PPM PCB;0.2 G TISSUE)
METHANE +CI, ~0.5 torr,105°

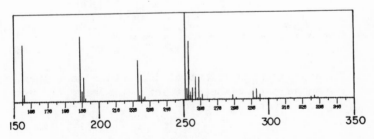

Figure 14. Positive methane chemical ionization mass spectrum
of human adipose tissue extract containing 1.2 ppm
total PCB residues.

CI AND NCI SPECTRA OF TISSUE EXTRACTS

Since the CI mass spectra of the Aroclor materials were relatively free of fragment ions, we decided to begin our studies of
direct CI examination of tissue extracts with human adipose tissue
samples that contained relatively large quantities of PCB residues.
Figs. 14 and 15 illustrate the CI and NCI mass spectra of partially
purified tissue extract which contained 1.2 ppm of a mixture of PCB
residues. The major component in the extract was p,p-dichlorobenzo-
phenone which is the chemical degradation product of the p,p'-DDT
and the DDT-type pesticide residue compounds that were originally
present in the extract. The p,p-dichlorobenzophenone appears as an
ion at m/e 251 (MH$^+$) in the CI spectrum and at m/e 250 (M$^{-\cdot}$) in the
NCI spectrum. The CI spectrum shows a series of MH$^+$ ions at m/e 155,
m/e 189, m/e 223, m/e 257, m/e 291 and m/e 325, which correspond to
biphenyl and its chlorinated derivatives through C15 .

The NCI mass spectrum of the tissue extract exhibited an ion
at m/e 231 which probably corresponds to p,p-dichlorobenzophenone
(Mole. Wt. 250) plus oxygen minus chlorine. The (M-19)$^-$ ion series
starts at m/e 271 and proceeds through Cl$_8$ at m/e 441. The ion group
at m/e 282 appears to correspond to a complex of the p,p-dichloro-
benzophenone molecule ion and oxygen (O$_2$).

TISSUE EXTRACT 1101 (1·2 PPM PCB;0·2 G TISSUE)
METHANE, -CI,~0.5 torr, 105°

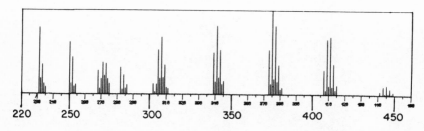

Figure 15. Negative methane chemical ionization mass spectrum of
            human adipose tissue extract containing 1.2 ppm total
            PCB residues.

SUMMARY

The CI and NCI mass spectra of polychlorinated pesticides are
generally simple, intense and easily interpreted. The two types of
spectra show different patterns for the same molecules so that they
can be used together to increase the analytical reliability in the

examination of mixtures obtained from natural sources. With few exceptions the CI spectra of polycyclic pesticides are dominated by $(M-Cl)^+$ ions while their NCI spectra are dominated by $(M+Cl)^-$ ions. The same is true of the aromatic DDT type pesticides. The CI spectra of DDE and other olefinic relatives show intense MH^+ or $M^{+\cdot}$ ions while their NCI spectra usually have $(M+Cl)^-$ as the base peak. The polychlorinated biphenyl compounds produce MH^+ ions exclusively under CI conditions. The NCI spectra are dominated by $(M-19)^-$ anions. These ions probably correspond to phenoxides which would be the result of a reaction of a molecular anion with oxygen (O_2).

The CI and NCI spectra of partially purified tissue extracts have clearly shown that these methods have sufficient sensitivity for direct application in screening sample extracts for pesticide residues at ppm levels and very probably at sub-ppm levels as well. Judging from the intensity of the spectra obtained for these samples it should be possible to detect specific residues at the ppb or ppt levels. Experiments to determine the sensitivity limits and the applicability of these methods to screening crude extracts are currently in progress.

REFERENCES

1). This work has been supported by a grant from the National Science Foundation and has been conducted under the partial auspices of the U. S. Environmental Protection Agency.

2). F. H. Field, Accounts Chem. Res. 1, 42 (1968).

3). G. W. A. Milne, H. M. Fales and T. Axenrod, Anal. Chem. 43, 1815 (1971).

4). R. C. Dougherty, J. D. Roberts and F. J. Biros, Unpublished Observations.

5). C. E. Melton, "Principles of Mass Spectrometry and Negative Ions," M. Dekker, New York, 1970.

6). H. P. Tannenbaum, J. D. Roberts and R. C. Dougherty, Unpublished Observations.

7). F. J. Biros, R. C. Dougherty and J. Dalton, Org. Mass Spectrom. 6, 1161 (1972).

8). R. C. Dougherty, J. Dalton and F. J. Biros, Org. Mass Spectrom. 6, 1171 (1972).

A SELF-TRAINING INTERPRETIVE AND RETRIEVAL SYSTEM FOR MASS SPECTRA. THE DATA BASE

F. W. McLafferty, M. A. Busch, Kain-Sze Kwok, B. A. Meyer, Gail Pesyna, R. C. Platt, Ikuo Sakai, J. W. Serum, Akira Tatematsu, Rengachari Venkataraghavan, R. G. Werth

Department of Chemistry, Cornell University, Ithaca, New York 14850

We have recently described[1,2] a self-training interpretive and retrieval system (STIRS) for computer interpretation of mass spectra which utilizes directly data of all available reference spectra, and does not require prior spectra/structure correlations of these data either by human or computer effort. The computer selects different classes of data known to have high structural significance, such as characteristic ions, series of ions, and masses of neutrals lost, from the unknown mass spectrum, and matches these against the corresponding data of all the reference spectra. The reference compounds of closest match in each data class are examined for common structural features; criteria have been determined so that such features can be identified with approximately 95% reliability. Each reference compound has been coded in Wiswesser Line Notation (WLN) to make possible computer recognition of structural features. Further details of the initial system are available.[1,2]

The original evaluation of STIRS utilized a reference file of 13,000 mass spectra.[1,2] These results showed clearly that the value of the interpretation by STIRS is directly dependent on the quality and quantity of the reference spectral data used. No human correlation of the spectral data is necessary; the addition of reference spectra of a new compound class immediately increases the capability of STIRS to interpret the spectra of such compounds. This realization provided the incentive for the program described here which has led to our present file of 25,000 reference mass spectra; the author list of this paper reflects the effort that

was involved in this project. We will first describe the sources
from which this quantity of data was assembled, and the steps taken
to insure the quality of these data. Then examples of spectral
interpretations utilizing this file and a new computer system will
be described.

SOURCES OF SPECTRAL DATA

The data were obtained from a wide variety of sources, as shown
in Table I.

Table I. Origin of File Data

Number	Source
6652	Atlas of Mass Spectral Data
6000	Aldermaston (AWRE, Aldermaston, England)
1073	Archives of Mass Spectral Data, Vols. 1, 2, and 3
497	Unpublished Archives Data
190	Drug Spectra, MIT Subcommittee 6 Collection (Prof. Biemann)
80	"Applications Tips" Finnigan Corp., Sunnyvale, Calif. (Environmental Pollutants)
400	Pollutants Spectra, Environmental Protection Agency, Athens, Georgia
68	Stanford Estrogen Spectra (Prof. Djerassi)
506	American Petroleum Institute Research Project 44 (College Station, Texas)
270	Thermodynamic Research Center Collection
150	Manufacturing Chemists Association Collection
9500	Reference Books and Journals (including A. Tatematsu and T. Tsuchiya, Ed., "Structural Indexed Literature of Organic Mass Spectra," Academic Press of Japan, Tokyo, 1966-8)
25386	Total

We are greatly indebted to many people for supplying and checking these spectra. There are approximately 18,000 different compounds represented in this reference file of over 25,000 spectra.

DATA QUALITY

The average spectrum in the collection contains over 100 peaks; the chances for error are thus relatively large. These errors arise from spectrometer performance, measurement inaccuracies, and data transcription. Checking for errors has been done in general by comparison to other mass spectra, either of the same or related compounds, by mass spectrometrists and by the computer.

All of the Atlas spectra and the first 4,000 of the Aldermaston spectra (Table I) were checked by mass spectrometrists against expected mass spectral behavior in this laboratory; the Archives spectra were each checked by two outside referees. It is presumed that many of the remaining spectra from the literature and other sources had been similarly checked before publication.

All of the spectra were also checked by a computer program for three common symptoms of poor data: (a) impurity peaks, (b) isotopic abundance ratios which do not correspond to those expected from the elemental composition, and (c) the presence of illogical neutral losses. Impurity peaks are defined as those with mass values greater than $(M + 3 + N)$ with relative abundances greater than 0.2% of the base peak. M is the mass of the molecular ion; the value of N is nonzero only when peaks above $(M + 3)$ can be reasonably expected (for example, if several atoms of Cl, Br, S, and/or Si are present).

To determine an error in the isotopic abundance ratios, the expected values of the $(M + 1)$ and $(M + 2)$ peaks are calculated and compared to the experimental values. The calculation considers contributions from the $(M - 1)$ and (M) peaks due to any single, naturally occurring isotope as well as any possible combination of two naturally occurring isotopes of the same or different elements (for example, two ^{13}C or one ^{13}C and one ^{2}H). A discrepancy of $\pm20\%$, or an absolute difference of 2.5/1000, whichever is greater, constitutes an error in the $(M + 1)$ or $(M + 2)$ peak.

Some small, primary neutral losses occur so rarely that a large peak representing one of these losses may be indicative of an error in the data. In the mass region above one half the molecular weight, the program searches for peaks at $(M - 2)$ through $(M - 13)$, $(M - 14)$ if $(M - 15)$ does not exist, $(M - 19)$ through $(M - 25)$, and $(M - 35)$ through $(M - 40)$. If one or more of these peaks is present above some threshold level, an illogical neutral loss error is indicated. Threshold levels are generally defined between 1 -- 5% of the base peak, depending on the specific neutral loss involved and how likely a logical occurrence of the peak is considered to be. If D, F, and/or Cl are present, (M - 2), (M - 19) and (M - 20), and (M - 35) and (M - 36), respectively, become logical. If Br or Cl is present, allowance is also made for isotopic contributions from logical peaks to peaks representing illogical losses.

Often it is more difficult to correct an error than to find it. If there is only one spectrum of the compound in the file, the mass spectrometrist may have to make an arbitrary decision on this. For example, he may have to guess as to the identity of the impurity giving anomalous peaks in the spectrum in order to subtract the total spectrum of the impurity from the reference spectrum. If the impurity cannot be identified, it is usually better to leave the impurity peak(s) in the spectrum as an indication of the spectral quality. It should be emphasized that such spectra are still very useful in the file, as STIRS has been found to be relatively insensitive to impurities.[1,2]

If more than one spectrum of a compound is in the file, selection is made using the "best spectrum program" (BSP). Spectra with none of the three types of errors described above are selected first. If all the spectra of a compound are found to contain errors, those with only illogical neutral losses are selected over those containing incorrect isotopic abundance ratios or impurity peaks. If all duplicate spectra also contain isotopic abundance errors, those without impurity peaks are selected. Finally, since most spectra are now run with ionizing potentials of 50 eV or higher, spectra taken at ionizing potentials less than 50 eV are excluded, whenever possible, by treating them as if they contained an error comparable to an impurity peak.

If BSP is unable to select a single, best spectrum from the error criteria, this is selected from the best set of spectra on the basis of: A) total number of peaks, B) mass discrimination, and C) source from which the data were obtained. This is expressed as an F factor, where $F = A^2 B C D$. A is defined as the total number of peaks in the spectrum, excluding those above (M + 3 + N). A threshold level of 0.2% of the base is set in order to omit spurious peaks due to noise and possible impurity peaks present below the molecular ion. The A term tends to exclude data in which the peaks are recorded over only part of the spectrum.

The B term tends to exclude data in which there is a large degree of mass discrimination in the low mass region. BSP begins with the molecular ion peak and searches the set of best spectra until it finds the highest mass peak meeting the following requirements: 1) it must occur at an amu value no greater than M, but not less than one half the molecular weight; 2) it cannot be one amu above the base peak; and 3) it must have a relative

abundance of at least 4% of the base in one spectrum and at least
1% of the base in all remaining best spectra to be compared; this
tends to eliminate the possibility that the B value is primarily
the result of noise or other experimental error. If such a peak
is found, B is defined as the absolute value of the peak abundance
relative to a base of 1,000. Whenever a peak cannot be found
which meets all three requirements, B is set to one.

The C term can be used to weight the selection in favor of
data that is known to be more reliable. For example, if part of
the reference file has been checked for errors (for example, the
"Archives" data), the C term can be defined as some number greater
than one for the more reliable data and equal to one for the
remaining reference file.

The D term can be used to discriminate against spectra taken
under unusual experimental conditions. As an example, spectra
taken at an ionizing potential less than 50 eV were assigned a D
value of 0.01, while all other spectra were assigned a D value
of one.

Squaring the A term creates a very strong preference for a
spectrum with the largest number of peaks, especially if the
compounds are of low molecular weight and only a few peaks are
present. However, as the total number of peaks in the spectrum
increases, the B term becomes relatively more important in controlling the F factor. For example, if C and D are both one, then
a spectrum with 399 peaks would require a B value of only 754 to
be chosen over a spectrum with 400 peaks and a B value of 750.
However, a spectrum with 9 peaks would require a B value of 926
to be chosen over a spectrum with 10 peaks and a B value of 750.
It was felt that this would be an advantage, since the omission
of data is more likely to be significant in spectra where the
total number of peaks is small, as compared to spectra where the
total number of peaks is very large.

A final feature of BSP compares spectra with the same WLN
to determine whether they are exact duplicates; for example,
some published spectra have also been submitted to the Aldermaston
file. An exact duplicate check is initiated whenever the difference
in the A values for two spectra is less than 3% and the ratio of
the two B values is 100% ± 2%. If both criteria are met, the
spectra are compared peak for peak. They are declared exact
duplicates if the number of mismatches does not exceed 7% (a
mismatch occurs when a peak is present in one spectrum and absent
in the other), and if the averaged sum of the ratios of the
intensities for all matched peaks is 100% ± 2%.

BSP was written for the IBM 360, model 65, and coded with
504 statements in PL1 machine language. Comparison of 10 duplicates

requires a maximum of 130 K of memory. Processing 170 sets of
duplicates in a previously sorted file of 25,000 compounds requires
an average of 1 minute cpu time. A printout for the program lists:
1) the spectra compared, 2) the errors found in each as well as
the amu values and relative abundances of the peaks involved,
3) the values determined for all A, B, C, D, and F terms, and
4) the final choice for the best spectrum along with other choices
allowing for any or all of the three types of spectrum errors.

WISWESSER LINE NOTATION SORTING PROGRAM (WSP)

For the examination of STIRS results by the chemist we find
that it is advantageous to have the spectra in the file arranged
to group related candidate structures. It was felt that such an
ordering was also desirable for the "Registry of Mass Spectral
Data."[3] Reference files are typically ordered by molecular weight
and elemental composition. However, whenever the presence of some
functional group is suspected, it is also convenient if the known
spectra are further arranged into groups of chemically related
compounds.

WSP arranges compounds of the same molecular weight and
elemental composition so that compounds with the same number and
kinds of functional groups appear together in the reference file.
Within each molecular weight-elemental composition batch, the
WLN symbols are first alphabetized. This causes all carbocyclic
and heterocyclic compounds to appear together since their WLNs
always begin with an "L" or "T", respectively; geometric isomers
also appear together as they have nearly identical WLNs except
for a few symbols near the end of the code. In addition, because
the compounds are ordered by nominal molecular weight, calculated
by substituting the mass of the most abundant isotope for any
isotopic label (for example, using an amu of one for deuterium),
then alphabetization of the WLNs also causes all compounds which
are identical except for elemental isotopes to be grouped together.

After alphabetization, WSP examines the WLN of each compound
for 34 selected symbol patterns and some of their permutations
(Table II). The priority of each symbol approximates the IUPAC
rules for choosing the "principal group" in a molecule. Thus a
compound with both a carboxylic acid and an amine group would be
placed near other carboxylic acids. As a result of WLN syntax,
locant symbols can be easily excluded since these are always
preceded by a space.

Pattern recognition is based on scanning the WLN for break
characters. The symbols surrounding the break characters are
examined to match the longest applicable pattern. Pattern
ambiguities are resolved on the basis of the left to right scan

Table II. Group Priorities for IUPAC[a] and WSP.

IUPAC Priority[b]	Formula	WLN Patterns	WSP Priority
Cations	X^+	e.g., &7/1	not coded except NR_4^+, 21
Carboxylic Acids	-COOH	VQ, QV	1
S Analogs	-CSOH	VSH, SHV	2
Sulfonic Acids, Sulfones	$-SO_3H$ $-SO_2-$	SW, WS	3
Sulfinic Acids, Sulfoxides	$-SO_2H$ $-SO-$	SO, OS	4
Anhydrides	$\overset{O}{\overset{\|}{-C}}-O-\overset{O}{\overset{\|}{C}}-$	VOV	5
Esters	-COOR	VO, OV, VHO	6
Thioesters	-COSR	VS, SV	7
Acid Halides	-COX	VG, VE, VF, VI	13; halogen not implemented
Primary Amides	$-CO-NH_2$	VZ, ZV	8
Secondary Amides	-CO-NH-	VM, MV, VHM	9
Tertiary Amides	-CO-N<	VN, NV, VHN	10
Hydrazides	-CO-N-N<	VNN, VNM, etc.	combinations of 9, 10, 19, 20
Imides	-CO-NH-CO-	VMV	9, 13
Amidines	$-C(=NH)-NH_2$	YZUM	18, 20, 25, 31
Nitriles, Isonitriles	$-C\equiv N$ $-N\equiv C$	CN, NC	11
Aldehydes	-CHO	VH, HV	12
Ketones	>C=O	V	13

Table II Continued

IUPAC Priority[b]	Formula	WLN Patterns	WSP Priority
Ketenes (Metal Carbonyls)[c]	$=C=O$ } $-C\equiv O$ }	CO, OC	14
Thioketones	$>C=S$	SU, US	15
Alcohols, Phenols	$-OH$	Q	16
Thiols	$-SH$	SH, HS	17
Hydroperoxides	$-OOH$	OQ, QO	16, 22
Primary Amines	$-NH_2$	Z	18
Secondary Amines	$-NH-$	M	19
Tertiary Amines	$-N<$	N	20
Imines	$=NH$	M	19
(Hydrogen Free N^+)[c]	$-\overset{\shortmid}{\underset{\shortmid}{N}}-^+$	K	21
Hydrazines	$-\overset{\shortmid}{N}-\overset{\shortmid}{N}-$	NN, NM, etc.	19, 20
Ethers	$-OR$	O	22
(Dioxo Group)[c]	$-X{\nearrow}^O_{\searrow O}$	W	23
Thioethers	$-SR$	S	24
(Double Bonds)[c]	$=$	U	25
(Triple Bonds)[c]	$\equiv$	UU	26
(Nitro Group)[c]	$-NO_2$	NW, WN	27
(Nitroso Group)[c]	$-NO$	NO, ON	28
(Benzene Ring)[c]	$-C_6H_5$	R	30
(Tertiary Carbon)[c]	$-\overset{\shortmid}{\underset{\shortmid}{C}}-$	X	31
(Secondary Carbon)[c]	$-\underset{\shortmid}{CH}-$	Y	32

Table II Continued

    a From Tables II and III, p. 10 and 12, <u>Nomenclature</u> <u>of</u> <u>Organic</u> <u>Chemistry</u>, Section C, IUPAC (1965).

    b Decreasing IUPAC Priority for Citation as a Principal Group.

    c Groups Enclosed in Parentheses Are not Included in the IUPAC List of Principal Groups.

---

rather than a hierarchial pattern matching algorithm. For simplicity, patterns were limited to lengths of three characters. Longer patterns are recognized in preference to shorter ones. Thus "VOVO" is given codes 5 and 22 instead of two code 6's.

This approach appears to work quite well for straight and branched-chain compounds. For heterocyclic compounds (beginning with "T") codes R, X, and Y are ignored. For carbocyclic compounds (beginning with "L") codes X and Y are ignored. This prevents chemically similar ring compounds from sorting into separate groups on the basis of branched side chains.

WSP is coded with 50 statements in Fortran IV and 321 statements in IBM 360 Assembly Language. Sorting a maximum of 47 WLN codes on an IBM 360/65 requires 46 K of memory. Processing 1350 alphabetized codes which have been previously ordered according to molecular weight and elemental composition requires approximately 1 minute of cpu time. A printout for the program lists the number of compounds sorted in each molecular weight-elemental composition batch, and the names and WLN codes of these compounds in order of decreasing pattern priority as recognized by the program.

RESULTS WITH EXPANDED DATA BASE

When STIRS compared the spectrum of 3-p-nitrophenylsydnone as an "unknown" against the original file of 13,000 reference spectra,[1] four of the first ten selections by the overall match factor (MF11) were aromatic sydnones. The improved performance with the new file of 25,000 spectra is shown in Table III.

Table III. Overall Match Factor Selections for 3-p-Nitrophenylsydnone

T5NNOVJ AR DNW

Compound	WLN	MF11
3-p-Bromophenylsydnone	T5NNOVJ AR DE	559
3-p-Fluorophenylsydnone	T5NNOVJ AR DF	516
3-(2-Naphthyl)sydnone	L66J C- AT5NNOVJ	511
3-p-Carboxyphenylsydnone	T5NNOVJ AR DVQ	504
3-m-Methoxyphenylsydnone	T5NNOVJ AR CO1	494
3-p-Methoxyphenylsydnone	T5NNOVJ AR DO1	494
3-p-(Carboxymethylphenyl)sydnone	T5NNOVJ AR D1VQ	481
3-o-Methoxyphenylsydnone	T5NNOVJ AR BO1	474
3-(4-Nitrophenyl)-5-oxazolidine	T5NMOV EHJ AR DNW	468
3-Phenyl-4-acetylsydnone	T5NNOVJ AR& EV1	467

The "Artificial Intelligence" program has recently been extended to the interpretation of high resolution spectra of estrogens. Professor Djerassi kindly supplied his estrogen reference data which was added to our reference file. The overall match factor (MF11) results for 17-vinylestradiol 3-methyl ether (which was excluded from the reference file for this search) are shown in Table IV. Note that for a more complete answer the chemist combines these MF11 results with molecular weight and functional group information (e.g., MF5, MF6, interactive shift technique).

Table IV. Overall Match Factor Selections for 17-Vinylestradiol
3-Methyl Ether, L E5 B666TTT&J BØ E F1U1 FQ IØ JØ OO1 -A&BI -B&EJ

Compound	MF11
17α-Ethinylestradiol 3-Methyl Ether	650
1,2-Dimethylestrone	517
Estradiol 3-Methyl Ether	512
6-Ketoestradiol	511
Estradiol 3-Methyl Ether 17γ-Lactone	501
1-Methylestrone 3-Methyl Ether	493
1-Methylestrone	477
Estrone 3-Methyl Ether	464
1-Methylestradiol	462
1-Methylestradiol 17-Monoacetate	439

These new STIRS results were run on a Digital Equipment
Corporation PDP-11/45 computer which has been operating in our
laboratory since April 1973. The condensed file of 25,000
reference spectra is searched on two 1.2M word removable head
disks. The present FORTRAN program requires approximately 12
minutes per unknown spectrum; we expect that the machine language
program now being implemented will be an order of magnitude faster.
This should expedite the more thorough testing and use of STIRS,
and the implementation of new match factors, such as new series
of neutral losses. We are also developing a method for computer
recognition of the common structural features from the WLN codes
of the selected references compounds for each match factor.
Finally, this faster program should also make possible a more
thorough check of each reference spectrum for errors by evaluating
it quantitatively with STIRS.

REFERENCES

(1) K.-S. Kwok, R. Venkataraghavan and F. W. McLafferty, J. Amer.
Chem. Soc., 95, June (1973).

(2)   Kain-Sze Kwok, Ph.D. Thesis, Cornell University, January 1973.

(3)   E. Stenhagen, S. Abrahamsson, and F. W. McLafferty, "Registry of Mass Spectral Data," Wiley-Interscience, New York City, 1974.

(4)   D. H. Smith, B. G. Buchanan, R. S. Engelmore, A. M. Duffield, A. Yeo, E. A. Feigenbaum, J. Lederberg, and C. Djerassi, J. Amer. Chem. Soc., <u>94</u>, 5962 (1972).

Acknowledgments

We are deeply indebted to the National Institutes of Health (PHS grant GM 16609) and the Environmental Protection Agency (grant R-801106) for generous support of this research program, and to Wiley-Interscience, Inc., for funds for collection of the spectral data.

Ion Kinetic Energy Spectroscopy: A New Mass Spectrometric Method

for the Unambiguous Identification of Isomeric Chlorophenols

S. Safe, W. D. Jamieson, and O. Hutzinger

Atlantic Regional Laboratory, National Research Council

of Canada, Halifax, Nova Scotia

Ion kinetic energy (IKE) spectroscopy is a technique which records the ionic decompositions which occur in the first field-free region of the mass spectrometer prior to the electric sector[1,2]. These ions can be recorded by suitable defocussing techniques, by scanning the electric sector voltage from 0 to 100% energy (E) the daughter ions pass through the β slit and these can be duly detected and recorded.

Electron impact spectra often do not distinguish between isomeric compounds such as o,p' and p,p' DDT and DDD[3] and most isomeric polychlorinated biphenyls (PCB's)[4]. It has recently been shown that the IKE spectra readily distinguish between isomeric DDT's[5], DDD's[5], PCB's[6], chlorobenzenes[6] and hexachlorocyclohexanes[5].

Chlorophenols are extensively used as fungicides and in addition are encountered in the environment as pesticide hydrolysis or degradation products. The electron impact spectra do not readily distinguish between all the isomeric chlorophenols[7] and we have therefore investigated the IKE spectra of these compounds to determine the analytical capabilities of this technique and to study decompositions occurring in the first field-free region of the mass spectrometer.

RESULTS AND DISCUSSION

Monochlorophenols

The ion kinetic energy spectra of the three isomeric

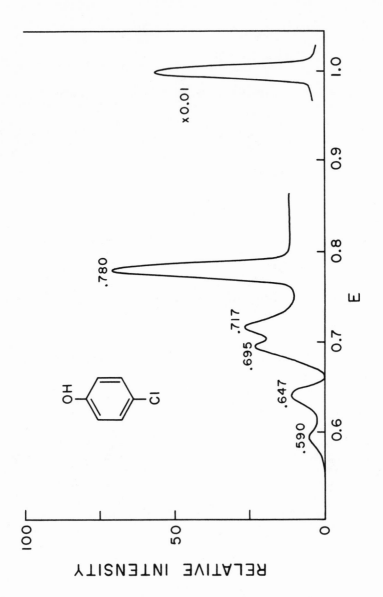

Figure 1; Ion kinetic energy spectrum of 4-chlorophenol

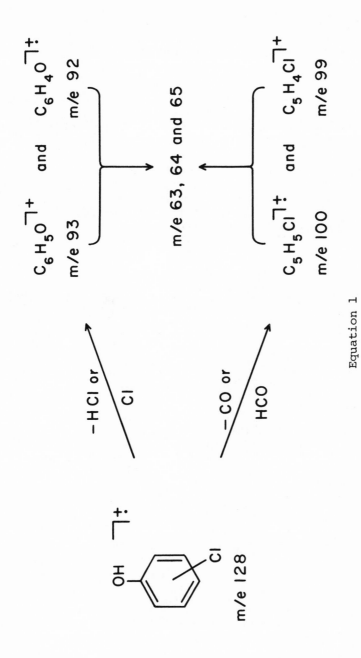

Equation 1

chlorophenols (i.e. Figure 1) exhibited peaks at 0.780, 0.717, 0.695, 0.647 and 0.590 E corresponding to the ionic decompositions $\underline{m/e}$ 128 → $\underline{m/e}$ 100 or $\underline{m/e}$ 99 (0.782 or 0.774 E), $\underline{m/e}$ 128 → $\underline{m/e}$ 93 or $\underline{m/e}$ 92 (0.723 or 0.716 E), $\underline{m/e}$ 100 → $\underline{m/e}$ 65 or $\underline{m/e}$ 64 (0.647 or 0.637 E) or $\underline{m/e}$ 99 → $\underline{m/e}$ 64 or $\underline{m/e}$ 63 (0.643 or 0.633 E), $\underline{m/e}$ 93 or $\underline{m/e}$ 92 → $\underline{m/e}$ 65 or $\underline{m/e}$ 64 (0.699 or 0.696 E) and $\underline{m/e}$ 65 or $\underline{m/e}$ 64 → $\underline{m/e}$ 39 or $\underline{m/e}$ 38 (loss of 26 mass units, 0.600 or 0.594 E). Due to the multiplicity of the fragmentation pathways (Equation 1) the ionic decomposition peaks were overlapping, however the IKE spectral data (Table 1) clearly distinguished between the three isomeric chlorophenols.

Dichlorophenols

The IKE spectra of the dichlorophenol isomers (i.e. Figure 2) give a distinctive fingerprint pattern for all of the dichloro isomers. The spectra gave ionic decomposition peaks at 0.825, 0.777, 0.735, 0.643 and 0.605 E corresponding to the reactions $\underline{m/e}$ 162 → $\underline{m/e}$ 134 (0.828 E), $\underline{m/e}$ 162 → $\underline{m/e}$ 126 (0.776 E), or $\underline{m/e}$ 126 → $\underline{m/e}$ 98 (0.774 E), $\underline{m/e}$ 134 → $\underline{m/e}$ 98 (0.729 E), $\underline{m/e}$ 98 → $\underline{m/e}$ 63 (0.729 E), and $\underline{m/e}$ 162 → $\underline{m/e}$ 98 (0.604 E) (see Equation 2). The results also indicated the elimination of the elements of HCOCl from the molecular ion ($\underline{m/e}$ 162 → $\underline{m/e}$ 98) also occurred in the first field-free region of the mass spectrometer and this reaction was not previously noted in the study of the primary ion mass spectra of these isomers.

It should be noted that the ionic decomposition peak for the expulsion of CO from the molecular ion from 2,6-dichlorophenol was observed in the IKE spectrum of this isomer although the ion at $\underline{m/e}$ 134 was not observed in the mass spectrum. Since the molecular ions decomposed with loss of CO and HCOCl the ratio of their ionic decomposition peaks [0.825 E]/[0.605 E] is a measure of the relative reactivity of the molecular ion and it was clear (Table 2) that this ratio was significantly different for most of the dichlorophenol isomers indicating retention of substituent identity in their decomposing molecular ions. The ratio is thus useful in

Table 1

Ion kinetic energy results for the isomeric chlorophenols

Compound	Peak abundances × 10^3				
	0.780 E	0.717 E	0.695 E	0.647 E	0.597 E
2-chlorophenol	14.9	21.0	17.3	2.1	1.6
3-chlorophenol	20.6	6.8	6.0	2.9	1.5
4-chlorophenol	24.4	6.4	4.7	3.7	1.7

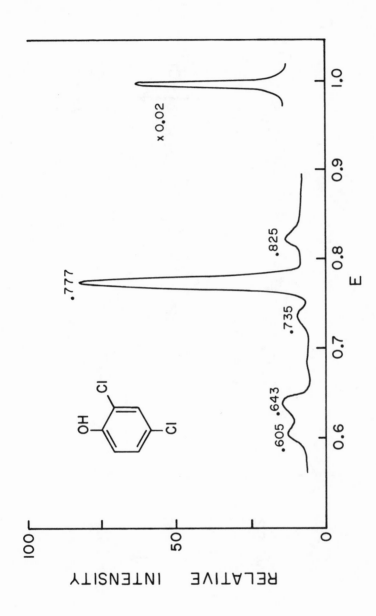

Figure 2; Ion kinetic energy spectrum of 2,4-dichlorophenol

$[C_6H_3O]^+$ m/e 91

$[C_5H_3Cl]^{+\cdot}$ m/e 98 $\xrightarrow{-Cl}$ $[C_5H_3]^+$ m/e 63

$-Cl\cdot$

$-CO$

HCl

$[C_6H_3ClO]^{+\cdot}$ m/e 126

$[C_5H_4Cl_2]^{+\cdot}$ m/e 134 $\xrightarrow{-Cl}$ $[C_5H_4Cl]^+$ m/e 99

$-HCl$

$-CO$

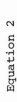

$[\text{dichlorophenol (OH, } Cl_2)]^{+\cdot}$ m/e 162

$\xrightarrow{-HCO}$ $C_5H_3Cl_2$ m/e 133

Equation 2

Table 2

Ion kinetic energy data for the isomeric dichlorophenols

Isomer	Peak abundances × 10^3					
	0.825 E	0.777 E	0.735 E	0.643 E	0.606 E	[0.825 E] / [0.605 E]
2,6	2.0	3.8	0.8	4.2	2.2	0.90
2,5	5.6	20.6	1.9	2.6	6.6	0.85
2,4	2.0	30.4	1.4	3.8	2.8	0.70
2,3	0.6	34.0	0.5	3.0	1.3	0.45
3,5	7.3	13.6	5.1	1.6	4.7	1.55
3,4	6.0	9.4	6.4	1.0	3.9	1.55

Table 3

Ion kinetic energy results for the isomeric trichlorophenols

Compound	Peak abundances × 10^3					
	0.857 E	0.817 E	0.777 E	0.727 E	0.675 E	[0.857 E] / [0.075 E]
2,3,4-	0.2	13.1	ca. 0.1	2.2	0.1	2.0
2,3,5-	0.7	28.8	1.0	3.6	0.7	1.0
2,3,6-	0.2	14.0	0.2	3.0	~0	–
3,4,5-	2.5	13.2	0.5	2.7	4.6	0.55
2,4,6-	0.2	17.0	0.2	2.7	0.6	0.30

distinguishing between the isomers.

Trichlorophenols

The IKE spectra (i.e. Figure 3) of the isomeric trichloro-phenols (Table 3) gave ionic decomposition peaks at 0.857, 0.817, 0.777, 0.727 and 0.675 E corresponding to the reactions m/e 196 → m/e 168 (0.858 E), m/e 196 → m/e 160 (0.815), m/e 161 → m/e 133 or m/e 132 (0.827 or 0.821 E), or m/e 160 → m/e 132 or m/e 131 (0.826 or 0.820 E), m/e 160 → m/e 125 (0.779 E), m/e 131 → m/e 96 (0.727 E) and m/e 196 → m/e 132 (0.673) (Equation 3).

IKE abundance data (Table 3) clearly showed some significant differences in the spectra of all the trichlorophenol isomers however the two peaks at 0.826 and 0.815 E were not completely resolved. In agreement with the mass spectra, the reaction

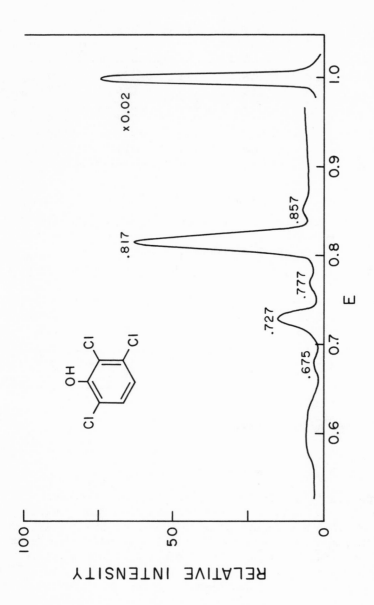

Figure 3: Ion kinetic energy spectum of 2,3,6-trichlorophenol

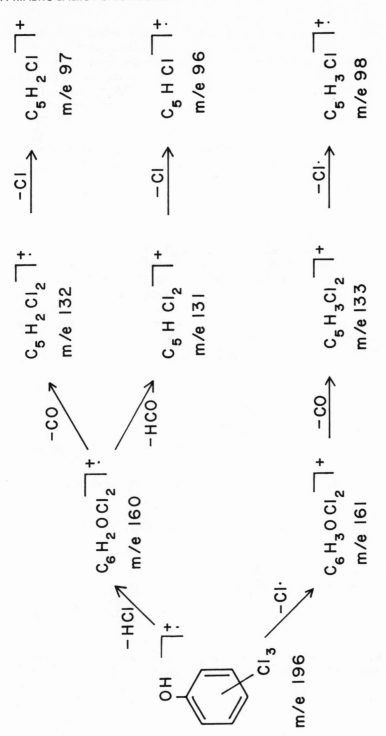

Equation 3

expelling CO from the molecular ion m/e 196 → m/e 168 was most intense for the 3,4,5-isomer and as with the dichlorophenols an ionic decomposition peak at 0.675 E was due to the expulsion of the element of formylchloride (HCOCl) from the molecular ion. Similarly the ratio [0.875 E]/[0.675 E] was different for all the isomers indicating energetically dissimilar decomposing molecular ions for all the trichlorophenol isomers.

The IKE method is clearly a useful technique for distinguishing between isomeric chlorophenols and is a further aid in the identification of these compounds.

EXPERIMENTAL

The chlorophenols were obtained from Aldrich and Eastman Chemical Co. and crystallized to constant melting point before use. The IKE spectra were recorded on a Dupont CEC 21-110B double-focussing mass spectrometer using a direct introduction wide range probe[8]. The IKE spectra were obtained by scanning the electric sector voltage as previously described and the data are the average results obtained for 4 to 5 scans.

REFERENCES

1.   Beynon, J. H., Baitinger, W. E. and Amy, J. W.   Int. J. Mass Spectrom. Ion Phys. 3, 47 (1969).

2.   Beynon, J. H. and Cooks, R. G.   Res/Develop. 22, 26 (1971).

3.   Sphon, J. A. and Damico, J. N.   Org. Mass Spectrom. 3, 51 (1970).

4.   Safe, S. and Hutzinger O.   J. Chem. Soc., Perkin Trans. 1, 686 (1972).

5.   Safe, S., Hutzinger, O., Jamieson, W. D. and Cook, M. Org. Mass Spectrom. 7, 217 (1973).

6.   Safe, S., Hutzinger, O. and Jamieson, W. D.   Org. Mass Spectrom. 7, 169 (1973).

7.   Biros, F. J. and Ross, R. T.   163rd National American Chemical Society Meeting, Boston, Mass., April (1972).

8.   Jamieson, W. D. and Mason, F. G.   Rev. Sci. Instr. 41, 778 (1970).

ELECTRON IMPACT SPECTRA OF SEVERAL AROMATIC SYSTEMS

A. Curley, R.W. Jennings, V.W. Burse, E.C. Villanueva

United States Environmental Protection Agency

Chamblee Toxicology Laboratory, Chamblee, Ga. 30341

Introduction: In the process of investigation of several commercial preparations and formulations of pesticides and pesticide related compounds for contaminants this laboratory observed interesting evidence of *ortho-para* substituent effects in aryl systems. The substituent of interest among these compounds is chlorine. The effect of the *ortho-para* aryl substituent was observed in the mass spectrometric fragmentation patterns. The compounds examined were phenyl ether, octachlorodiphenyl ether, 2,7-dichlorodibenzo-*p*-dioxin, 2,3,7,8-tetrachlorodibenzo-*p*-dioxin, octachlorodibenzofuran, octachlorodibenzo-*p*-dioxin and dibenzofuran.

Experimental: The compounds were examined on the LKB Gas-Chromatograph-Mass Spectrometer at 70 eV ionization energy. All compounds except octachlorodiphenyl ether were analyzed by direct inlet. Source-analyzer conditions were: Filament current-3.8A, Trap current-60 µA, Box current-40 µA, Leak current-5 µA, Analyzer pressure-10-7mm, Analyzer temperature-290o C, Probe temperature 30o C, Box and Source temperature 290° C. Phenyl ether was supplied by Fisher Scientific, Fairlawn, New Jersey; dibenzofuran was supplied by Aldrich Chemical Co., Milwaukee, Wis.; the dichloro-dibenzo-*p*-dioxin and octachlorofuran were supplied by Analabs, Inc., North Haven, Conn.; and the tetrachlorodibenzo-*p*-dioxin was supplied by Dow Chemical, Midland, Mich. All compounds assayed at 99% purity. Octachlorodiphenyl ether was extracted from pentachlorophenol and purified by the method of Firestone et al.[1] and examined on 3% SE-30, 10' X 1/4", 80/100 Chromosorb W, A.W., DMCS treated, at 230°C. The inlet temperature was 245°C.

Results and Discussion: Plate 1[*], shows the fragmentation of

---

[*]Ring hydrogen atoms are not illustrated in drawings.

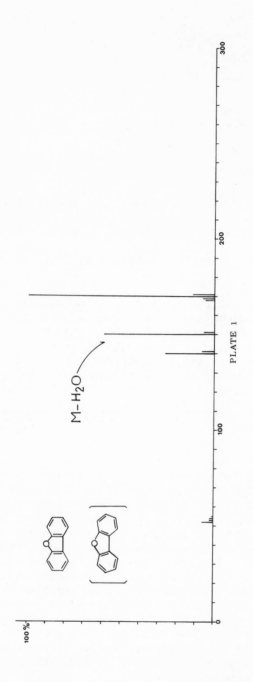

dibenzofuran. The most striking fragment is M-H_2O at M^+/e = 150 with intensity of 58% of base peak. The concerted loss of 18 mass units competes favorably with the two-step loss of 29 mass units, M-(H + CO). The absence of fragments below M^+/e = 139 supports the conclusion that double ring closure must be occurring to yield a cyclopentadiene-phenylene system described by McLafferty[2]. The intensity of the M^+/e = 50 probably involves an αH transfer to the ring with retention of charge on the opened ring fragment.

Octachlorodibenzofuran, Plate 2, yields the expected spectrum with no evidence of fragmentation below M^+/e = 185. The parent M^+/e = 440 shows losses of one and two chlorine atoms with M-2Cl being more pronounced. A more significant pattern develops from M-Cl. From evidence gathered in other studies, the M-Cl leading to the loss of CO must involve a chlorine atom *ortho* (α) to the oxygen bearing carbon atom. M-(Cl + CO), M^+/e = 377, loses one chlorine to yield M^+/e = 342, and two chlorines to yield M^+/e = 307 with M-(3Cl + CO) being intense. The spectrum reveals doubly charged ions at M^{++}/e = 220, M^{++}/e = 188.5, and M^{++}/e = 185 with isotopic clusters corresponding to the isotopic contributions of the singly charged counterparts at M^+/e = 440, 377, and 370, respectively.

Phenyl ether, Plate 3, has been described[3][4]in the literature. The spectrum does not contain necessarily unexpected features. The appearance of an unexpected base peak in the higher chlorinated diphenyl ethers requires an explanation of the absence of a stronger M-2 fragment, or M^+/e = 139, which occurs with dibenzofuran. The low probability of homolytic expulsion of the o and o' hydrogen atoms is probably related to the random C-H bond cleavage around the rings and an increase in the α C-H bond strength through 2 p orbital electron resonance. The diphenyl ether does not close to the furan or does not acquire the furan isoelectronic structure as does the octachlorodiphenyl ether. Plate 4, octachlorodiphenyl ether, yields principle fragments at M-2Cl, the base peak (cluster), and M^+/e = 201, a Cl_4 cluster. The M^+/e = 201 is probably the tetrachloropentadienyl radical ion. No further fragmentation appears at the level of sample enrichment of our analytical runs.

Plate 5, 2,7-dichlorodibenzo-*p*-dioxin, shows a stronger ion at M^+/e = 188, M-(HCl + CO), than at M-35. Strong fragmentation below M^+/e = 150 appears to indicate degradation not seen in any of the other compounds examined except the diphenyl ether[5]. M^+/e 160 = M-(Cl + H + 2CO). A strong ion at M^+/e = 126 shows an absence of Cl^{37} isotopic patterns precluding the chlorophenoxy radical ion and is probably the doubly charged parent ion. M^+/e = 160 - Cl accounts for ion M^+/e = 125 of low intensity. Below M^+/e = 78 is a series of ions reminiscent of the aromatic nucleus in the process of degrading.

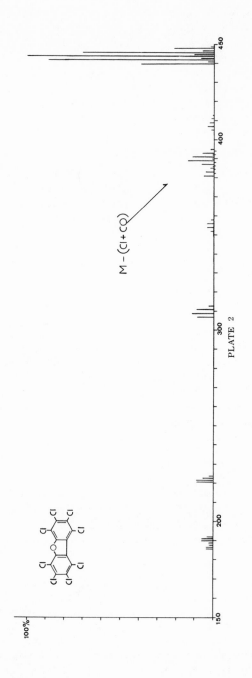

PLATE 2

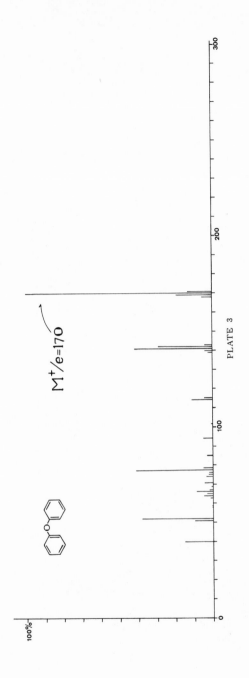

M+/e=170

PLATE 3

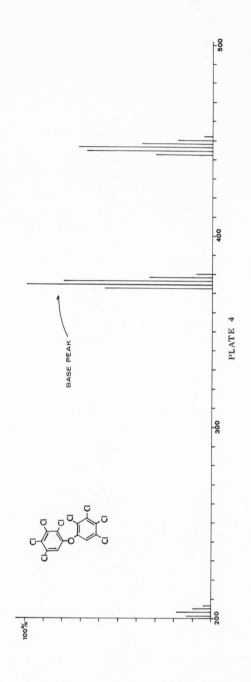

PLATE 4

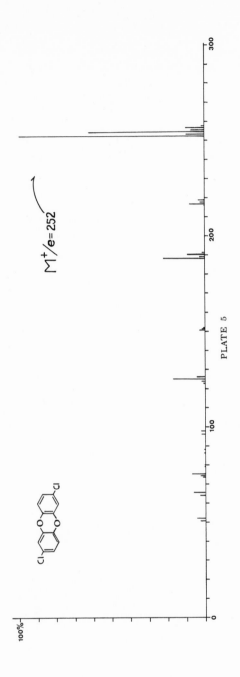

M$^+$/e=252

PLATE 5

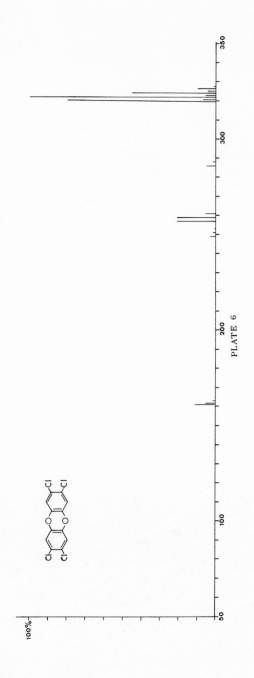

PLATE 6

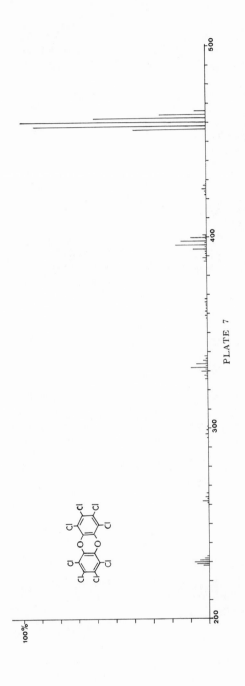

PLATE 7

Plate 6,[5] 2,3,7,8-tetrachlorodibenzo-p-dioxin, shows a base
peak in the parent ion isotopic cluster. Principle fragmentation
is M-35 and M-(35 + 28). A minor M-(2Cl + H) appears at M^+/e = 249.
A strong doubly charged parent ion, M^{++}/e, appears at mass 160.

Plate 7,[5] octachlorodibenzo-p-dioxin, shows a base peak at
M^+/e = 456 with the strongest secondary ion appearing at M^+/e = 393,
M-(Cl + CO). An ion at M^+/e = 386 is M-2Cl. M^+/e = 330, M-(2CO +
2Cl), is very intense. An ion at M^+/e = 228 is apparently the
doubly charged ion.

Conclusion and Extensions: The tendency toward modulation of
the skeletal fragmentation patterns is quite apparent in the group
of compounds studied. The surprising loss of water in the dibenzofuran is definitely modified by the presence of chlorine in the
molecule. The gross differences in the fragmentation of the
diphenyl ether and the chlorinated diphenyl ether indicate that
the presence of chlorine in the aromatic ring system greatly alters
the tendency of some skeletons of molecules to fragment along the
paths followed by the skeletons of their unsubstituted counterparts.

The group of compounds examined here is being extended to include other compounds and isomers which may establish the existence
of a general trend which we have attempted to characterize. This
group of compounds includes known position isomers of the chloro-
diphenyl ethers and the chlorodibenzofurans (2-chlorodiphenyl ether,
3-chlorodiphenyl ether, 2,2'-dichlorodiphenyl ether, 3,3'-dichloro-
diphenyl ether, 2,8-dichlorodibenzo-p-dioxin, 2,8-dichlorodibenzo-
furan and 3,7-dichlorodibenzofuran). These compounds are in the
process of being acquired through purchase or through synthesis
and purification. Many of these compounds and their stereoisomers
have been observed in extracts and their spectra examined, but the
compounds have not been fully characterized.

References:
1. Firestone, D., Ress, J., Brown, N., Barron, R., and Damico, J.:
   J.A.O.A.C. 55, No. 1, p. 85, 1972.
2. McLafferty, F.: Mass Spectrometry of Organic Ions, Academic
   Press Inc., N.Y., N.Y. p. 374, 1963.
3. Budzikiewica, H., Djerassi, C., and Williams, D.: Interpretation
   of Mass Spectra of Organic Compounds, Holden-Day, San Francisco,
   Calif., p. 181-182, 1965.
4. Beynon, J., Lester, R., and Williams, A.: J. Chem. Phys. 63,
   p. 1861, 1959.
5. Plimmer, J., Ruth, J., and Woolson, E.: J. Ag. and Food Chem. 21,
   p. 90, 1973.

IDENTIFICATION OF INSECTICIDE PHOTOPRODUCTS BY MASS
SPECTROMETRY

Earl G. Alley and Bobby R. Layton

Mississippi State Chemical Laboratory

Mirex (dodecachloropentacyclo[$5.3.0.0^{2,6}.0^{3,9}.0^{4,8}$]decane),
1, has been used extensively in the southeastern United States to
control the imported fire ants (<u>Solenopsis invicta</u> and <u>Solenopsis
richteri</u>). The irradiation of this chlorocarbon in hydrocarbon
solvents with ultraviolet light produced $C_{10}Cl_{11}H$ and $C_{10}Cl_{10}H_2$
derivatives (Alley et al., 1973). Three isomers are possible,
2, 3, and 4. The infrared spectrum of the $C_{10}Cl_{11}H$ photoproduct
differed from that reported by Dilling et al. (1967) for 1,2,3,4,
5,5,6,7,8,9,10-undecachloro[$5.3.0.0^{2,6}.0^{3,9}.0^{4,8}$]decane (2).
Nothing was known about the chemical reactivity of the different
chlorine atoms in Mirex; therefore, synthesis of compounds with
known structures was not possible. Furthermore, neither infrared,
mass spectral, nor nmr data provided any information to help
establish which of the remaining isomers (3 or 4) was the photoproduct.

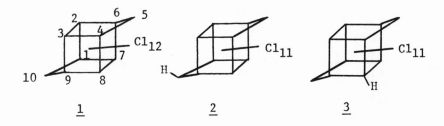

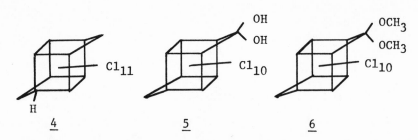

This investigation has produced evidence that Kepone hydrate, $\underline{5}$, (decachloro-5,5-dihydroxypentacyclo[5.3.0.0^{2,6}.0^{3,9}.0^{4,8}]-decane) and the dimethyl ketal of Kepone, $\underline{6}$, (decachloro-5,5-dimethoxypentacyclo[5.3.0.0^{2,6}.0^{3,9}.0^{4,8}]decane) were also photochemically reduced. The assignments of the structures of the photoproducts of these compounds were made from their mass spectra. The structure of the $C_{10}Cl_{11}H$ photoproduct of Mirex could then be assigned because chemical conversion of the primary photoproducts of Kepone derivatives to their corresponding Mirex analogues (Equation 1) produced compounds identical to the Mirex photoproduct.

The mass spectra of several pentacyclic compounds with the Mirex skeleton have been reported (Dilling and Dilling, 1967). The major fragmentation modes were dechlorination and cleavage of the carbon skeleton into two cyclopentadiene rings (Equation 2).

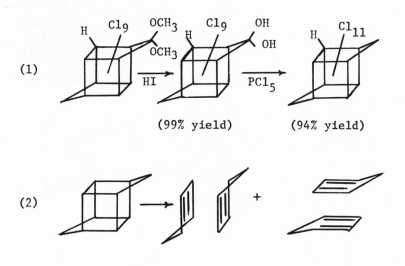

For Mirex, the mass spectrum had peaks corresponding to $C_{10}Cl_{12}^+$, $C_{10}Cl_{11}^+$, $C_5Cl_6^+$, and $C_5Cl_5^+$ ion clusters, along with other ions. Within an isotopic cluster, the number of lines and their intensities are determined by the number of chlorine atoms present and the natural abundance of the chlorine isotopes; therefore, the assignment of the number of chlorines contributing to an ion cluster is unambiguous. In this study the observed number of lines and their relative intensities within a cluster were generally in excellent agreement with those predicted from the molecular formula assigned to the ion. The only exceptions were in the case of the ketal derivatives, where two or more fragments had the same mass but different molecular formulas.

The mass spectrum of the $C_{10}Cl_{11}H$ photoproduct of Mirex (Figure 1a) indicated that the fragmentation pathways of the photoproduct were the same as those for Mirex. Dechlorination of the parent ions gave an ion cluster for $C_{10}Cl_{10}H^+$ ions; whereas, the ion clusters for the cyclopentadiene fragments were assigned the molecular formulas C_5Cl_6, C_5Cl_5H, and C_5Cl_5. The mass spectral fragmentation patterns of all the $C_{10}Cl_{11}H$ isomers should be identical (Table I); consequently, no structural assignment can be made from their mass spectra.

Table I

C_5 Cleavage Fragments of the Mirex Derivatives

Compound	Position of the Hydrogens (See Structure 1)	Predicted Cyclopentadiene Fragments		
		C_5Cl_6	C_5Cl_5H	$C_5Cl_4H_2$
$C_{10}Cl_{11}H$	10	+*	+	−
	9	+	+	−
	8	+	+	−
	observed for photoproduct	+	+	−
$C_{10}Cl_{10}H_2$	(10,6),(10,5),(9,6),(9,4),(8,2),(8,3)	−	+	−
	(10,10),(10,9),(9,1),(8,7)	+	−	+
	(10,8),(9,7),(9,8),(8,6)	+	+	+
	observed for photoproduct	−	+	−

* + present
  − not present

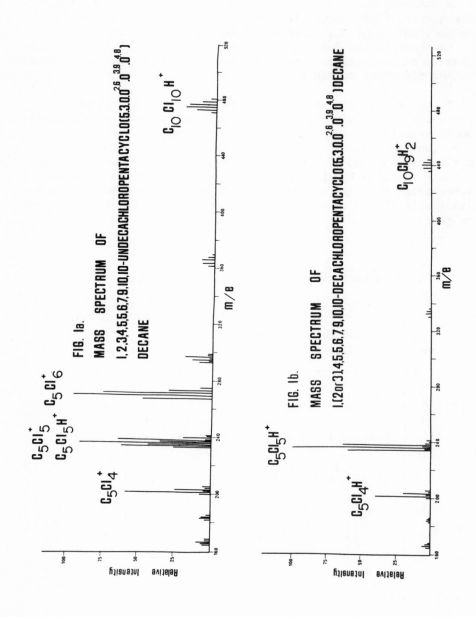

The mass spectrum of the $C_{10}Cl_{10}H_2$ photoproduct (Figure 1b) had major peaks for $C_5Cl_5H^+$ ions, but none for $C_5Cl_4H_2^+$ or $C_5Cl_6^+$ ions. The structures consistent with the mass spectral data (Table I) were ones with hydrogens on the following positions: (8 and 2), (8 and 3), (9 and 6), (9 and 4), (10 and 5), or (10 and 6). Dilling et al. (1967) reported that a hydrogen on the position alpha to a chlorine was shielded less (δ 4.3) than hydrogens on other positions (δ 3.74) in Mirex derivatives. Since the nmr spectra for both monohydro and dihydro photoproduct had one sharp peak at δ 3.7, the only structures for the dihydro photoproduct which would fit all the data have hydrogens at positions (8 and 2), (8 and 3), (9 and 6), or (9 and 4).

In order to obtain more information concerning the photochemistry of Mirex, the reactions of similar pentacyclic compounds were investigated. Gas-liquid chromatographic (glc) analysis indicated that the ultraviolet irradiation of Kepone hydrate, 5, in cyclohexane gave two major products. Attempts to isolate these photoproducts by glc and liquid-solid adsorption chromatography were unsuccessful. However, the mass spectra of these compounds were obtained with a mass spectrometer interfaced with a gas chromatograph. Dehydration occurred during the gas chromatographic separation of Kepone hydrate and its photoproducts. Consequently, the spectra obtained were for the ketones rather than the gem-diols. The mass spectral analysis indicated that two photoproducts were formed with molecular formulas $C_{10}Cl_9H_3O_2$ and $C_{10}Cl_8H_4O_2$. A mixture of the photoproducts of Kepone hydrate was converted to the corresponding Mirex derivatives by treatment with phosphorus pentachloride. The resulting Mirex derivatives were isolated by liquid-solid adsorption chromatography. The nmr and infrared spectra of these Mirex derivatives were identical to the corresponding Mirex photoproducts.

The mass spectral fragmentation modes of Kepone (decachloropentacyclo[5.3.0.0^{2,6}.0^{3,9}.0^{4,8}]decan-5-one) are analogous to those for Mirex (Dilling and Dilling, 1967). However, Kepone is less symmetrical than Mirex; and when the pentacyclic skeleton of Kepone is cleaved, C_5Cl_6 and C_5Cl_4O fragments result. The mass spectrum (Figure 2a) of the $C_{10}Cl_9HO$ derivative of Kepone had major peaks for $C_{10}Cl_8HO^+$ ions (dechlorination of the parent); $C_5Cl_6^+$ and $C_5Cl_5H^+$ ions (fragments from the cleavage of the pentacyclic carbon skeleton); and $C_5Cl_5^+$ ions (dechlorination of the C_5Cl_6 fragment). Of the four possible $C_{10}Cl_9HO$ isomers of Kepone, only structure 7 (1,2,3,4,6,7,9,10,10-nonachloropentacyclo-[5.3.0.0^{2,6}.0^{3,9}.0^{4,8}]decan-5-one) could give both C_5Cl_6 and C_5Cl_5H fragments from the cleavage of the carbon skeleton (Table II).

The mass spectrum(Figure 2b) of the Kepone derivative ($C_{10}Cl_8H_2O$) consisted of major peaks for $C_{10}Cl_7H_2O^+$ and $C_5Cl_5H^+$ ion clusters, but none for $C_5Cl_6^+$ or $C_5Cl_4H_2^+$ ions. The observed fragmentation pattern could be obtained only from isomers 8a, 8b, 9a, 9b, or 10.

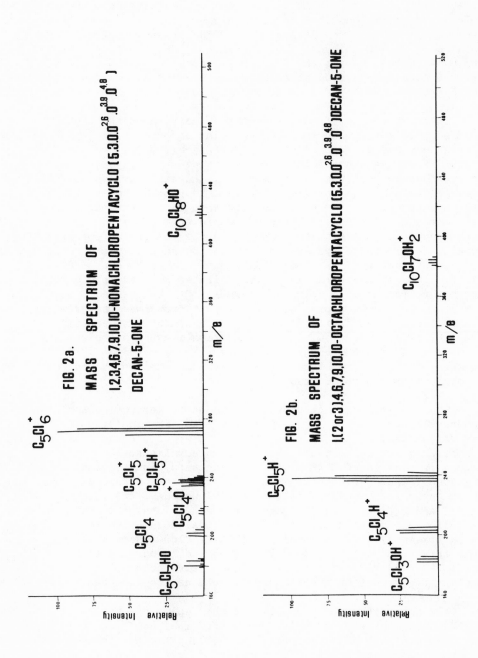

FIG. 2a.

MASS SPECTRUM OF

1,2,3,4,6,7,9,10,10-NONACHLOROPENTACYCLO (5.3.0.0$^{2.6}$.0$^{3.9}$.0$^{4.8}$)

DECAN-5-ONE

FIG. 2b.

MASS SPECTRUM OF

1,[2 or 3],4,6,7,9,10,10-OGTACHLOROPENTACYCLO (5.3.0.0$^{2.6}$.0$^{3.9}$.0$^{4.8}$)DECAN-5-ONE

Table II

C5 Cleavage Fragments of the Kepone Compounds

Compound	Positions of the Hydrogens (See Structure 7)	Predicted Cyclopentadiene and Cyclopentadienone Fragments					
		C_5Cl_6	C_5Cl_5H	$C_5Cl_4H_2$	C_5Cl_4O	C_5Cl_3HO	$C_5Cl_2H_2O$
$C_{10}Cl_9HO$	10	-*	+	-	+	-	-
	9	-	+	-	+	-	-
	8	+	+	-	+	+	-
	6	+	-	-	-	+	-
	observed for dehydrated photoproduct	+	+	-	+	+	-
$C_{10}Cl_8H_2O$	(6,4)	+	-	-	-	-	+
	(10,6),(9,6),(9,4),(8,2),(8,3)	-	+	-	-	+	-
	(10,10),(10,9),(9,1)	-	-	+	+	-	-
	(8,6), (8,4)	+	+	-	-	+	+
	(8,7)	+	-	+	+	-	+
	(10,8),(9,8),(9,7)	-	+	+	+	+	-
	observed for dehydrated photoproduct	-	+	-	-	+	-

+ present
- not present
*

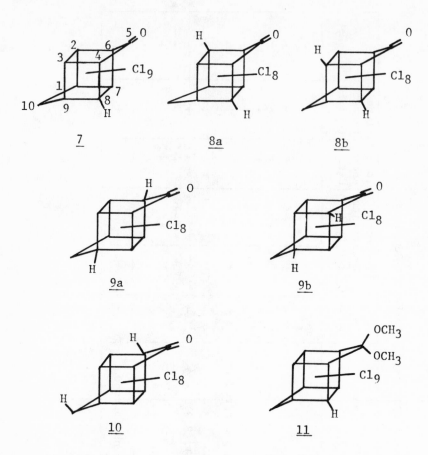

Since the photoproducts of Kepone hydrate could not be isolated, another derivative of Kepone (the dimethyl ketal) was included in the photolytic study. The photolysis of this compound
yielded two products with molecular formulas $C_{12}Cl_9H_7O_2$ and
$C_{12}Cl_8H_8O_2$. As expected, the mass spectrum of the $C_{12}Cl_9H_7O_2$ photoproduct had peaks for both $C_5Cl_6^+$ and $C_5Cl_5H^+$ ions. Based on
arguments similar to those used for the Kepone derivative ($C_{10}Cl_9HO$),
this photoproduct was assigned structure 11 (1,2,3,4,6,7,9,10,10-
nonachloro-5,5-dimethoxypentacyclo[5.3.0.0^2,^{6}.0^3,^{9}.0^4,^{8}]decane).
The infrared spectrum of compound 11, isolated by liquid-solid chromatography, did not have a band at 1600 cm^{-1} (C=C). The nmr spectrum of this compound had three sharp singlets: δ 3.56(1), δ 3.54(3),
and δ 3.51(3). Of the possible isomers, only compound 11 would be

expected to have different chemical shifts for the two methyl
groups. When photoproduct $\underline{11}$ was treated with hydriodic acid in
glacial acetic acid, the corresponding Kepone hydrate derivative
was formed. The mass spectrum and glc retention time of this
compound were identical to those for the $C_{10}Cl_9H_3O_2$ photoproduct
of Kepone hydrate. Treatment of this compound with phosphorus
pentachloride gave a substance whose infrared spectrum was identical to the one for the Mirex photoproduct.

The mass spectrum (Figure 2b) of the ketal $C_{12}Cl_8H_8O_2$ photoproduct had peaks for $C_5Cl_5H^+$ ions but none for $C_5Cl_6^+$ or $C_5Cl_4H_2^+$
ions. Ketals with structures analogous to $\underline{8a}$, $\underline{8b}$, $\underline{9a}$, $\underline{9b}$, and
$\underline{10}$ are the only possibilities consistent with this data (see
Table II). Also, the nmr spectrum of this photoproduct had one
slightly broadened peak at δ 3.56; therefore, structure $\underline{10}$ may be
eliminated from consideration. The infrared spectra of the
$C_{10}Cl_{10}H_2$ photoproduct of Mirex and the Mirex derivative synthesized from the ketal photoproduct, $C_{12}Cl_8H_8O_2$, were identical.

The preceding arguments unequivocally established the assign-
ment of structure $\underline{3}$ to the $C_{10}Cl_{11}H$ photoproduct of Mirex. Since
compound $\underline{3}$ is a precursor of the $C_{10}Cl_{10}H_2$ photoproduct of Mirex
(Alley et al., 1973), the possible structures for this photo-
product are $\underline{12a}$ and $\underline{12b}$. Photoproducts of both Kepone hydrate
and the dimethyl ketal of Kepone can be converted to this same
dihydro photoproduct of Mirex. Therefore, the possible structures
for the octachloro photoproducts are $\underline{13a}$ or $\underline{13b}$ for the hydrate
and $\underline{14a}$ and $\underline{14b}$ for the ketal.

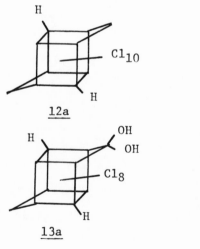

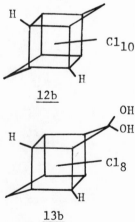

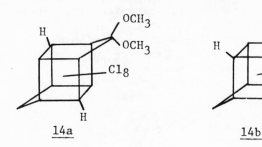

14a   14b

Acknowledgement:

    The authors are indebted to the Entomology Research Division
of the United States Department of Agriculture for financial support for this research.

References:

E. G. Alley, D. A. Dollar, B. R. Layton, J. P. Minyard, Jr., J.
    Agr. Food Chem., 21, 138 (1973).
W. L. Dilling, H. P. Braendlin, E. T. McBee, Tetrahedron, 23,
    1211 (1967).
W. L. Dilling and M. L. Dilling, Tetrahedron, 23, 1225 (1967).

MASS FRAGMENTOGRAPHY AS APPLIED TO SOME ORGANOPHOSPHATE INSECTICIDES

Joseph D. Rosen and Stephen R. Pareles

Department of Food Science

Rutgers University, New Brunswick, N.J. 08903

INTRODUCTION

Analysis of pesticides in environmental samples poses difficult problems for the chemist because he is looking for the proverbial needle in the haystack. Although the use of gas-liquid chromatography in conjunction with sensitive detectors specific to certain atoms has been of considerable help, there are many samples which require rigorous cleanup to prevent interference with the analysis. Furthermore, there are many instances where the absolute presence of the material under analysis is open to question because there is not even enough material for a mass spectral confirmation. A relatively new technique, mass fragmentography, shows considerable promise as a sensitive, selective, and confirmative means of handling picogram quantities of pesticides (as well as other environmental pollutants).

EXPLANATION OF MASS FRAGMENTOGRAPHY

Mass fragmentography is a technique in which the mass spectrometer is employed as a detector for materials eluting from a gas chromatograph. Its advantages over other methods of detection are high

sensitivity and high selectivity, the selectivity not
being dependent on the presence of electron capturing
atoms or certain heteroatoms. In order to best understand how mass fragmentography works, let us first
consider the operation of a combined gas chromatograph-
mass spectrometer. Material eluting from the gas
chromatograph is rapidly conducted to the ion source
of the mass spectrometer where it is ionized into a
number of characteristic fragments. These fragments,
which have different mass to charge ratios are then
separated by the magnet and detected. Because the
detection has to be rapid, only a small fraction of
each fragment produced in the ion source is recorded.
For example, if we scan a material eluting from a gas
chromatograph between m/e 444 and m/e 44 at 4 seconds
per decade, the entire mass spectrum is complete in
4 seconds. This means that (assuming a liner scan for
purposes of simplified arithmetic) the mass spectrometer uses only .01 second to determine a fragment at
any one atomic mass unit. If we were interested in a
particular fragment most of the ions corresponding to
this fragment would never reach the detector, because
the mass spectrometer is busy detecting the other ions
formed for the remaining 3.99 seconds of the spectrum.
However, if instead of scanning, we were to set the
detector so that only the fragment of interest is detected, the entire time of elution of the material from
the gas chromatograph is spent detecting (and record-
ing) this fragment, leading to a large increase in
sensitivity. A further sensitivity increase comes from
being able to operate the oscillograph recorder at
lower response frequencies because rapid scanning is no
longer necessary. The ability to record at a lower
response frequency allows one to set the electron
multiplier gain at a higher setting. An alternative to
operating at very high electron multiplier settings is
to substitute a 1 mv potentiometric recorder for the
100 mv oscillograph recorder. This is an advantageous
alternative in that it allows one to operate at much
slower chart speeds, prolongs the life of the electron
multiplier, and saves on the cost of u.v.-sensitive
paper.

The end result of this is a mass fragmentogram

whose appearance is similar to that of a gas chromatogram, with the retention time of the material still
being dependent on the gas chromatographic conditions,
but the detection being dependent solely on the presence
of the fragment or fragments of interest.

Middleditch and Desiderio (1973) determined that
single ion detection was about 2 orders of magnitude
more sensitive than total ion detection. Total ion
detection is usually about 1 order of magnitude more
sensitive than flame ionization detection.

HISTORY OF MASS FRAGMENTOGRAPHY

Mass fragmentography was first proposed by Hammar
et al. (1968) who used it for the detection of the drug,
chlorpromazine, in human plasma. In addition to being
able to detect as little as one picogram of material,
the authors employed a device which permitted rapid
switching between three pre-set accelerating voltages,
allowing them to record the signal from three different
fragments simultaneously. Since the integrated areas
of the three peaks were proportional to the relative
intensities of the three fragments as determined earlier
by "normal" mass spectroscopy, they were able to ascertain that the mass fragmentograms were due only to
chlorpromazine. This early model had a range limited
to only 10% of the lowest mass under investigation, e.g.
from m/e 160 to m/e 176. Gaffney et al. (1971) reported
a lower limit of sensitivity of 50 picograms for
nortriptylene. Further refinements of the technique
were made by the development of the Multiple Ion Detector which could detect three different masses
simultaneously within a range of 20% of the lowest mass.
In the presence of interfering polychlorinated biphenyls
(PCB's), Bonelli (1972) was able to quantitate DDE in
the low nanogram region by focusing on m/e 246. Wolf
et al. (1972) reported a lower limit of sensitivity of
ca. 5 pg for chromium trifluoroacetylactone. Five 20
pg. on-column injections gave a relative standard
deviation of 6.2%. Strong and Atkinson (1972) used
mass fragmentography to monitor lidocaine and its

desethylated metabolite. They pointed out the advantage
of the technique on a quadrupole mass spectrometer where
there is no limit to the range of fragments that can be
simultaneously monitored. Presently, multiple ion
detection with magnetic instruments has advanced to a
range limited by 1.6 times the smallest fragment detected. Sjoquist and Anggard (1972) determined
homovanilic acid in cerebrospiral fluid. Although the
lower limit of sensitivity was not reported, it was
claimed that sensitivity was comparable to electron-
capture and the method faster and easier. Bieber et al.
(1972), by addition of a deuterated internal standard,
were able to quantitate the juvenile hormone present
in a single male Hyalophora cecropia. Sensitivity of
the order of 0.5 picomole was obtained by Koslow et al.
(1972) for the determination of pentafluoroacetyl
derivatives of norephinephrine and dopamine.

EXPERIMENTAL

A duPont 21-490 Mass Spectrometer interfaced to a
Varian 2740 Gas Chromatograph via a glass jet separator
was used. Operating parameters were a 4 ft. by 1/8 in.
glass column packed with 3% Dexil GC 300 on Anakrom
ABS, 80-90 mesh; carrier gas, Mathieson UHP helium
(ca. 30 ml/min); injection port and detector temperature,
245°C; molecular separator, 250°C; ion source, 290°C;
electron energy, 70 ev; filament current, 250 microamps.
Column temperatures of 140, 200, 210, and 225° C. gave
the following respective retention times: DDVP - 2.52,
diazinon - 3.85, malathion - 3.50, sumithion - 4.15
minutes.

In order to reject high frequency noise and provide additional zero correction to the 1 mv potentiometric recorder (Sargent Model SR), the amplifier output
of the mass spectrometer was followed by a VCVS low
pass filter network (single-pole, Butterworth) with
offset null control. The unit was built with a 741
operational amplifier, an Analoge Devices Model 902
Power Supply (± 15 volts), two 4700 ohm resistors and
two 15 microfarrad capacitors to give a cutoff frequency
of 2.2 Hz. as described by Tobey et al. (1971). In

addition, a 10-turn 10 kiloohm potentiometer was used
as a null control.

RESULTS AND DISCUSSION

The most abundant peaks found in a mass spectrum
of diazinon are found in Table I.

TABLE I

Major Diazinon Fragments

m/e	Relative Intensity
304	15
199	28
179	52
152	66
137	100
124	24
97	36
93	50

As m/e 137 was the most abundant ion, we set our mass
marker on 137 and optimized the peak with the accelerating voltage. Setting the electron multiplier at the
2 setting, we were able to get a straight line calibration curve for on-column injections between 8 ng. to
500 pg. The 500 pg on-column injection gave a 6.5%
scale deflection with a signal to noise ratio of 12.
By setting the electron multiplier at higher values,
we were able to easily detect 125 pg. of diazinon but
the signal to noise ratio was unacceptable for quantitative analysis. The single most serious problem
encountered in quantitating material was inability to
keep focused on the m/e 137 maximum. There were times
when we obtained reproducible results on injections as
much as 40 minutes apart but there were other times
when reproducibility could not be attained even after
5 minutes. We were able to overcome this problem with
some success by taking advantage of the column bleed.
After the 137 peak was optimized and the diazinon used
to optimize the peak was pumped away, the scale deflection due solely to column bleed was determined.
Any deviation from this level was then adjusted for by

adjusting the accelerating voltage. In this way we
were able to obtain an 8.7% relative standard deviation
for 5 successive 2 ng on-column injections. No impair-
ment to quantitation at this level was observed even
when 2 ng of diazinon were injected in a solution containing 5 micrograms spinach extract. No attempt was
made at clean-up of the spinach in order to demonstrate
how well the method works without sample clean-up.

Holland et al. (1973) have recently developed a
computer system for continuous fine adjustment of the
accelerating voltage in order to maintain optimal
focusing; thus making mass fragmentography, in our
opinion, the best method for the confirmative determination of picogram quantities of environmental
contaminants.

The ideal way to confirm the identity of picogram
amounts of material is with a multiple ion detector.
However, even without one, confirmation is still
possible. From Table I it can be seen that the relative
intensities of m/e 152 and 179 are approximately 2/3
and 1/2 respectively that of m/e 137. Since we were
able to easily distinguish 125 pg. of diazinon from
noise at m/e 137, we can theoretically confirm the
presence of ca. 200 pg of diazinon by focusing on m/e
152 and ca. 250 pg by focusing on m/e 179. While it is
impossible to state with what degree of certainty one
can ascribe the presence of these three peaks to
diazinon, some calculations at this point are instruc-
tive. First we assume that the 17,124 compounds listed
in "Eight-peak Index of Mass Spectra" (1970) are
representative of all the compounds in the universe and
then we count the number of compounds that exhibit mass
spectra that have among their eight most relatively
abundant ions those at 137, 152, and 179. The values
are 270, 345 and 244, respectively. This means that
the chance for any other compound to give a response
at all of these three mass settings is given by the
expression:

$$\frac{270}{17,124} \text{x} \frac{345}{17,124} \text{x} \frac{244}{17,124} = \frac{1}{220,848}$$

Making the further assumption that only the eight major peaks of an interfering material will contribute reasonably large signals, the odds are 220,847 to 1 that picogram quantities of diazinon can be confirmed by this method. These odds are even better when one considers the fact that g.c. retention time has not been considered in these calculations.

A material which gives a large number of ions of high relative intensity is good for confirmation but adversely affects the ultimate sensitivity. Diazinon, in a sense, is not the best compound for mass fragmentography because upon fragmentation there is no one ion which is produced in much greater yield than a number of others. Since the sensitivity of the mass fragmentography technique depends on the detection of a single ion, the sensitivity is reduced. An intriguing possibility to increase sensitivity, therefore, is to combine mass fragmentography with chemical ionization.

Other organophosphate insecticides that were studied were sumithion, malathion, and DDVP. All gave satisfactory calibration curves down to 4, 8, and 24 ng, respectively. The poor sensitivity of DDVP was probably due to decomposition on the stainless steel splitter. The lower limits for reliable quantitative analyses of these three materials are somewhat lower as the values for these three insecticides were obtained before installation of the VCVS low pass filter network.

LITERATURE CITED

M. A. Bieber, C. C. Sweely, D. J. Faulkner, M. R. Petersen, Anal. Biochem. 47, 264 (1972).

E. J. Bonelli, Anal. Chem. 44, 603 (1972).

"Eight Peak Index of Mass Spectra", Vol. 2, 1st Edition, Mass Spectrometry Data Centra, Berkshire, U.K. (1970).

T. E. Gaffney, C.-G. Hammar, B. Holmstedt, R. E. McMahon, Anal. Chem. 43, 307 (1971).

C.-G. Hammar, B. Holmstedt, R. Ryhage, Anal. Biochem. 25, 532 (1968).

J. F. Holland, C. C. Sweely, R. E. Thrush, R. E. Teets, M. A. Bieber, Anal. Chem. 45, 308 (1973).

S. H. Koslow, F. Cattabeni, E. Costa, Science 176, 177 (1972).

B. S. Middleditch and D. M. Desiderio, Anal. Chem. 45, 806 (1973).

B. Sjoquist, E. Anggard, ibid 44, 2297 (1972).

J. M. Strong, A. J. Atkinson, ibid 44, 2287 (1972).

G. E. Tobey, J. G. Graeme, L. P. Huelsman, "Operational Amplifiers, Design and Applications", McGraw-Hill, New York (1971), p. 297.

W. R. Wolf, M. L. Taylor, B. M. Hughes, T. O. Tiernan, R. E. Sievers, Anal. Chem. 44, 616 (1972).

ACKNOWLEDGEMENT

This research was supported, in part, by Regional Project NE-83, U. S. Department of Agriculture. Paper of the Journal Series, Agricultural Experiment Station, Rutgers, The State University of New Jersey, New Brunswick, N. J.

THE APPLICATION OF MASS SPECTROMETRY TO THE STUDY OF NITROANILINE-DERIVED HERBICIDES

Jack R. Plimmer and Ute I. Klingebiel

Agricultural Environmental Quality Institute

Agricultural Research Center-West, USDA, Beltsville, Md.

INTRODUCTION

Substituted dinitroaniline herbicides are increasing in economic importance. For example, trifluralin (α,α,α-trifluoro-2,6-dinitro-N-N-dipropyl-p-toluidine) is used extensively on soybeans and cotton in the southeastern, midwestern, and southern areas of the United States. Estimated production volume in 1971 was 25 million pounds compared with 90 million pounds of atrazine [4-chloro-2-(ethylamino)-6-(isopropylamino)-s-triazine and 45 million pounds of 2,4-D (2,4-dichlorophenoxyacetic acid). These two were the only herbicides to exceed the production volume of trifluralin. Related dinitroaniline herbicides are in commercial production and several members of this class are being developed. The structural formulae of a number of dinitroaniline herbicides are shown in Figure 1.

In view of their acceptance as herbicides, it has become important to elucidate the pathways by which dinitroanilines are metabolized or altered in the environment. Mass spectrometry has become recognized as an important technique for structural study and is extremely useful for identification or characterization, when limited quantities of material are available. Dinitroaniline herbicides, their photoproducts and metabolites often give complex mass spectra. We have examined some of the principal features of their fragmentation pathways and compared these with model compounds.

A brief summary of the known photochemical and biochemical transformations of the dinitroaniline herbicides provides an indication of the chemical characteristics of the structural types to be studied.

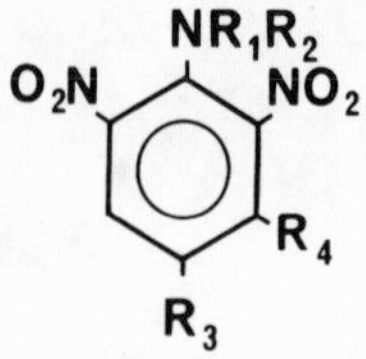

$$\begin{array}{c} NR_1R_2 \\ O_2N \diagdown \diagup NO_2 \\ \bigcirc \\ \diagdown R_4 \\ R_3 \end{array}$$

	R_1	R_2	R_3	R_4
Trifluralin	n-Pr	n-Pr	CF_3	H
Benefin	n-Bu	Et	CF_3	H
Dinitramine	Et	Et	CF_3	NH_2
Isopropalin	n-Pr	n-Pr	$CH(CH_3)_2$	H
Nitralin	n-Pr	n-Pr	SO_2CH_3	H
Oryzalin	n-Pr	n-Pr	SO_2NH_2	H
N-sec-buty-4-tert-butyl-2,6-dinitroaniline	sec-Bu	H	$C(CH_3)_3$	H
N,N-bis(2-chloroethyl)-2,6-dinitro-p-toluidine	$ClCH_2CH_2-$	$ClCH_2CH_2-$	CH_3	H

Structural Formulae of Dinitroaniline Herbicides

Figure 1

PHOTOCHEMICAL DEGRADATION

The photochemical reduction of nitroarenes has been known since the turn of the century when Ciamician and Silber (1905) studied many photochemical reactions, including the reduction of nitrobenzene to aniline. Among the products that may be obtained in the photoreduction are bimolecular compounds such as azoxybenzene. A reaction scheme (Figure 2) was proposed by Barltrop and Bunce (1968). In their discussion, they account for the wavelength dependence of the reaction products. At longer wavelengths, azoxybenzene and azobenzene were formed in greater yields. The reaction was initiated by excitation of nitrobenzene to the lowest triplet state $n\pi^*$ (Trotter and Testa, 1968), which then abstracted hydrogen from the solvent to give a radical ($Ar\dot{N}O_2$). Subsequently, this radical was reduced to aniline, via nitrosobenzene and phenylhydroxylamine, in a sequence of reactions that involved thermal as well as photochemical steps. Phenylhydroxylamine is the source of azo- and azoxybenzene. However, the ability of the excited nitroarene triplet state to abstract hydrogen was recently questioned (Davidson et al., 1971).

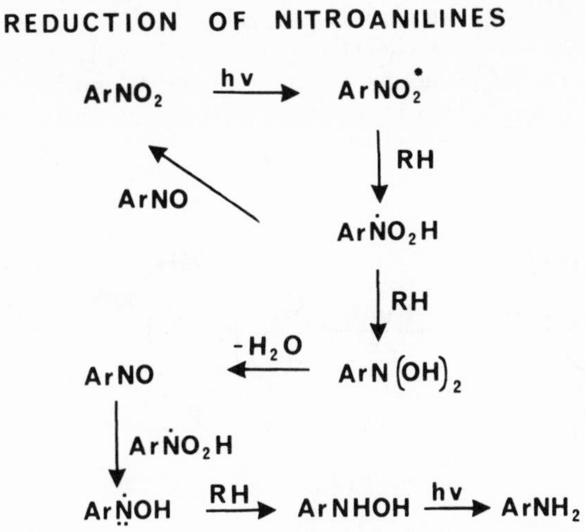
REDUCTION OF NITROANILINES
ArNO2
hv
ArNO2•
RH
ArNO
ArṄO2H
RH
-H2O
ArNO
ArN(OH)2
ArṄO2H
ArṄOH
RH
ArNHOH
hv
ArNH2

Figure 2

Reaction by way of a charge-transfer complex was suggested as an alternative route for photochemical reduction of the nitro group, accompanied by dealkylation that occurs in o-nitro-N-alkylanilines. Under mildly alkaline conditions N-(2,4-dinitrophenyl)leucine, or other 2,4-dinitrophenylamino acids, photodecomposed to 4-nitro-2-nitrosoaniline (Russell, 1963, 1964). The reaction was pH dependent and at low pH, the product was a benzimidazole N-oxide (Aloisi et al., 1972). The generality of this photochemical reaction was explored (Fielden et al., 1970; Preston and Tennant, 1972) and mechanisms were proposed (Meth-Cohn, 1970) for the degradation of N-2,4-dinitrophenyl α-amino acids and for the formation of benzimidazoles or their N-oxides (Fielden et al., 1970). Thus the acid-catalyzed photolysis of N,N-dialkylanilines gives either a benzimidazole or a benzimidazole N-oxide (Figure 3). At high pH a nitroso derivative is formed by the loss of N-alkyl groups.

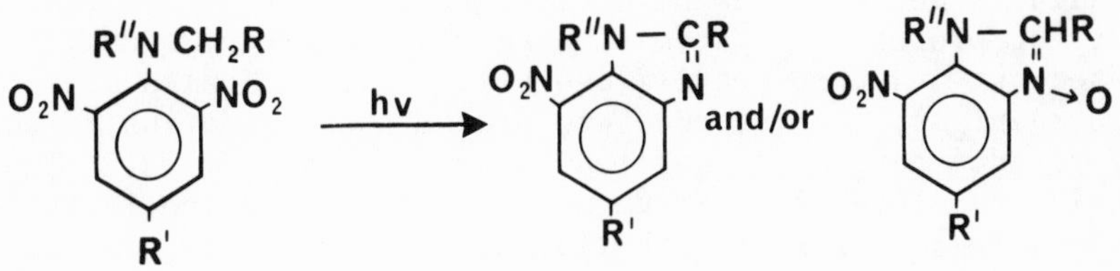
R″N CH2R
O2N
NO2
R′
hv
R″N — CR
O2N
N
R′
and/or
R″N — CHR
O2N
N→O
R′

Figure 3

Both mono-N-methyl and mono-N-propyl derivatives of α,α,α-trifluoromethyl-2,6-dinitro-p-toluidine (McMahon, 1966) on irradiation in heptane for 3 hours with a 450-watt mercury lamp gave α,α,α-trifluoro-2-nitro-6-nitrosotoluidine and the corresponding aldehyde (Figure 4).

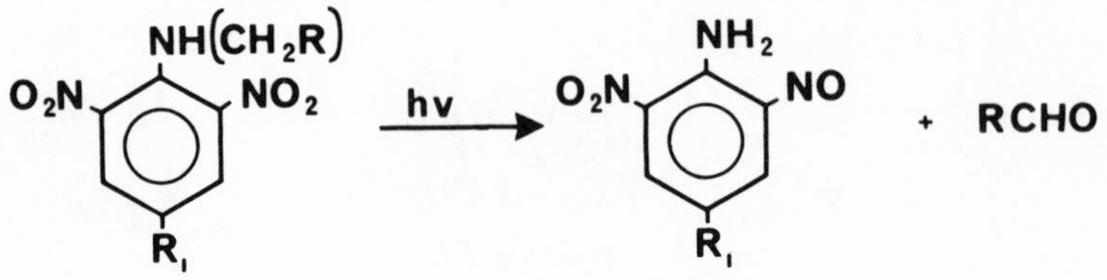

Figure 4

Although there was extensive decomposition of trifluralin on irradiation in methanol or heptane at neutral pH, monodealkylated and didealkylated products were identified (Day, 1969). Vapor phase photolysis afforded the monodealkylated product and a second compound tentatively postulated as having a benzimidazole structure (Crosby and Moilanen, 1972). A benzimidazole was detected among the products of irradiation of trifluralin on a thin-layer chromatographic plate (Golab, 1972).

Dinitramine (N^4,N^4-diethyl-α,α,α-trifluoro-3,5-dinitrotoluene-2,4-diamine) was rapidly converted on irradiation in water or methanol to a mixture of benzimidazoles and 1,2-dihydroxybenzimidazoline (Newsom and Woods, 1973). The formation of a benzimidazole by irradiation of N-sec-butyl-4-tert-butyl-2,6-dinitroaniline is prevented by the sec-butyl group but mass spectrometric evidence was obtained for the presence of cyclic compounds, m/e 251 and m/e 279, in trace quantities (Figure 5). Dealkylation and reduction of a nitro to a nitroso group are predominant photochemical reactions of this compound (Plimmer and Klingebiel, 1972).

METABOLIC PATHWAYS

Metabolic reactions parallel the photochemical pathways. Trifluralin is dealkylated and/or its nitro groups are reduced in a variety of biological systems (Kearney and Kaufman, 1969) (Figure 6). N-sec-butyl-4-tert-butyl-2,6-dinitroaniline is also dealkylated by microorganisms, and intermediate products of oxidation of the N-alkyl residue have been isolated (Kearney et al., 1972). The principal metabolite of dinitramine in soil was 6-amino-1-ethyl-2-methyl-7-nitro-5-(trifluoromethyl)benzimidazole (Smith, 1972). This compound was also identified as a metabolite of dinitramine in cultures of Paecilomyces and other soil fungi (Laanio et al., 1972) (Figure 7).

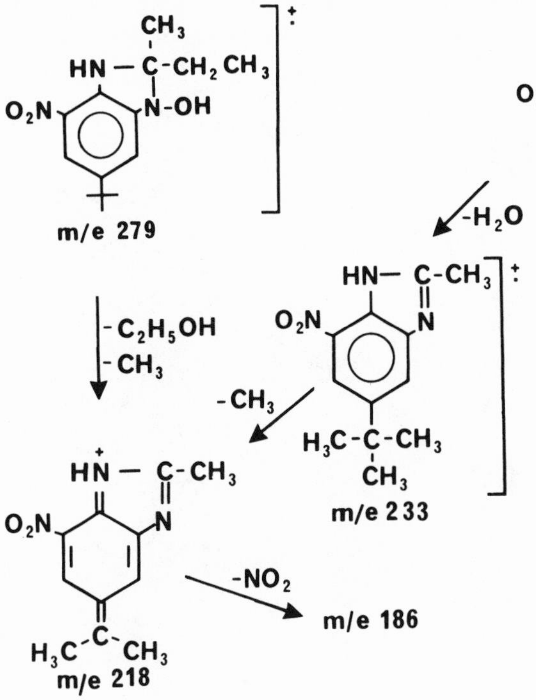
m/e 279
m/e 251
-H$_2$O
$-C_2H_5OH$
$-CH_3$
$-CH_3$
m/e 233
-NO$_2$
m/e 186
m/e 218

Figure 5

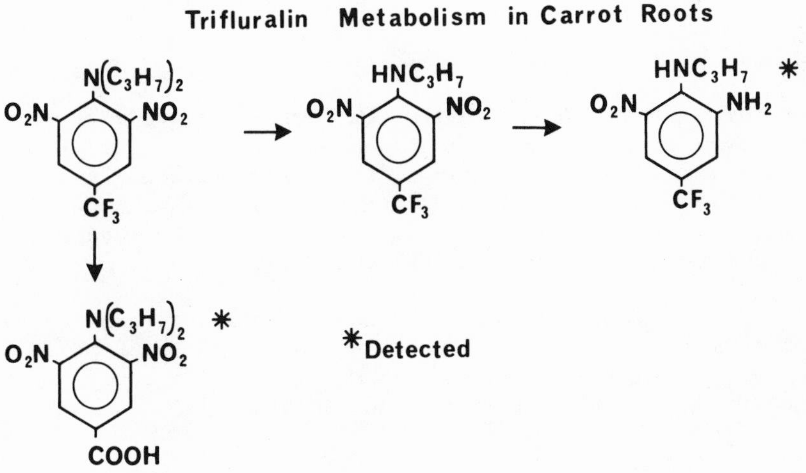
Trifluralin Metabolism in Carrot Roots
*Detected

Figure 6

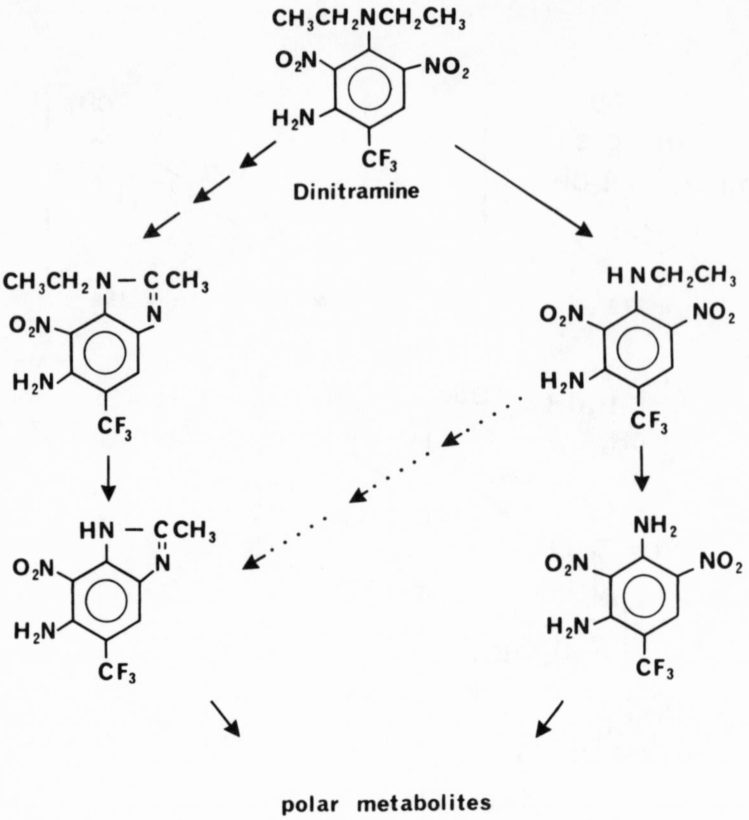

polar metabolites

Figure 7

EXPERIMENTAL

The compounds discussed in the text were obtained from several sources: purchased through commercial channels, from our previously described experimental work (Plimmer and Klingebiel, 1972) or donated by the following pesticide manufacturers: Amchem Products, Inc., Ambler, Pennsylvania; Eli Lilly and Company, Indianapolis, Indiana; and The Ansul Company, Weslaco, Texas.

Low resolution spectra were obtained with a Perkin-Elmer Model 270 GC/mass spectrometer via a probe and subjected to 80 EV. GC mass spectra were obtained using a 15-m open tubular capillary column (0.051 cm i.d., coated with SE-30 on Chromosorb W).

The high resolution spectra were determined by Dr. F. Biros of the Primate Laboratory in Perrine, Florida or obtained commercially. The samples were introduced via a probe and subjected to 70 EV in a Varian CH$_5$ double-focusing mass spectrometer.

MASS SPECTROMETRY: RESULTS AND DISCUSSION

Mass spectrometry permits the recognition of many characteristic features of the parent dinitroanilines and their metabolites. Low and high resolution spectra of a variety of substituted derivatives were determined to attempt to obtain correlations that might be useful in further structural investigations.

Substituted nitrobenzenes generally fragment by loss of NO from the molecular ion. Steric and electronic factors influence the ratio of the intensity of the molecular ion to that of the [M-NO]+ ion in the case of meta and para substituents (Bursey, 1969).

Relatively simple aromatic dinitro compounds behave in a similar manner to the mononitro analogs. For example, the fragmentation of 2,4-dinitrobenzene shows ions M^+, $[M-O]^+$, $[M-NO_2]^+$, $[M-NO_2-O]^+$, $[M-NO_2-NO]^+$, and $[M-2NO_2]^+$, and 2,6-dinitrotoluene loses two hydroxy groups (Budzikiewicz et al., 1967). This decomposition pathway is altered by the presence of ortho substituents bearing hydrogen atoms. The effects of interacting ortho substituents on electron-impact induced fragmentation is well documented and their recognition is valuable for identification of structural features. For example, the characteristic $[M-X]^+$ ions (where X = I, Br, Cl, OCH_3 or OH) shown by ortho substituted phenylhydrazones and 2,4-dinitrophenyl-hydrazones permit the ortho substituted derivatives to be distinguished from meta and para isomers (Cable et al., 1972).

Similarly, intense $[M-Cl]^+$ ions are characteristic of ortho-chlorophenylcarbamate (Still et al., 1972), -phenylthiourea and -phenylurea (Baldwin et al., 1967) fragmentation pathways. The elimination of water molecules was observed in the case of ortho-hydroxyacetanilide during electron-impact induced decomposition. This parallels the facile thermal decomposition by this route (Plimmer and Klingebiel, 1972). Hydrogen bearing ortho groups such as CH_3 can interact with the nitro group (Budzikiewicz et al., 1967). Elimination of OH and H_2O permits ortho nitroanilines to be distinguished from the meta and para substituted derivatives (Figure 8). However, in spite of the value of this process for recognition of the ortho relationship between nitro and amine substituents, its absence does not indicate that such a relationship is nonexistent. Other processes may dominate the fragmentation pathway. For example, the introduction of a para tertiary butyl group into 2,6-dinitroaniline suppresses the appearance of the $[M-OH]^+$ ion. The relative abundance of the $[M-OH]^+$ ion is 9.9 in 2,6-dinitroaniline and the base peak of the spectrum corresponds to the molecular ion. By contrast, 4-tert-butyl-2,6-dinitroaniline (I) fragments by loss of a methyl radical to give an ion $[M-15]^+$ which forms the base peak (Figure 9). This ion may be stabilized as a quinoid structure (II). Fragment ions $[M-61]^+$ and $[M-107]^+$ represent the loss of NO_2 and $2NO_2$ from the base peak.

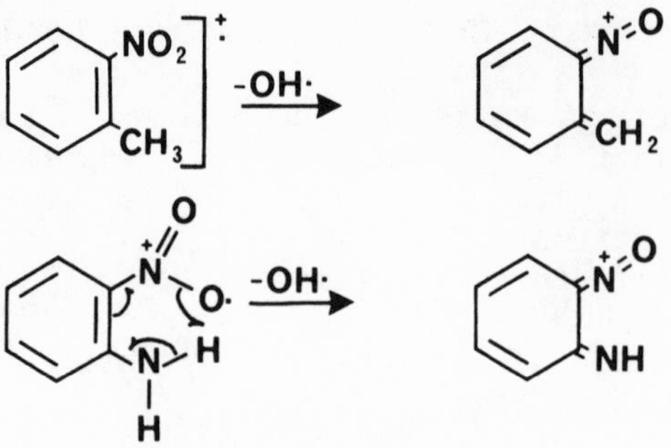

Figure 8

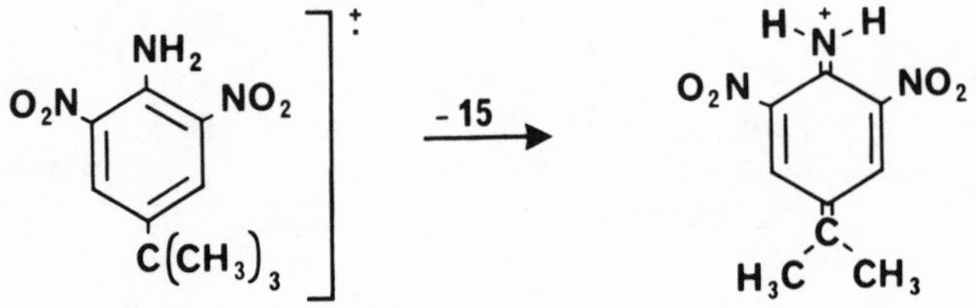

Figure 9

Fragmentation of 4-tert-butyl-2-nitro-6-nitrosoaniline (III) obtained by photolysis of the herbicide N-sec-butyl-4-tert-butyl-2,6-dinitroaniline, resembles that of (I) (Figure 9). The molecular ion is at m/e 223, and a peak at [M-15] corresponds to the loss of CH$_3$. A quinoid structure (Budzikiewicz et al., 1967) may again be postulated for this stable ion. High resolution measurements show that the ion at m/e 194 corresponds to the loss of CHO from the molecular ion. This unusual decomposition pathway has been noted previously in nitrosobenzenes (Schroll et al., 1968). Loss of CO from the molecular ion of 1-nitronaphthalene has also been noted as an atypical reaction (Budzikiewicz et al., 1967). Loss of CH$_3$ from (III) is accompanied by expulsion of NO (m/e 178) and (NO$_2$+O) (m/e 146). The base peak of the spectrum (m/e 132) corresponds to (IV), obtained by loss of (CH$_3$+NO+NO$_2$) from the molecular ion (Figure 10). This is also present as an intense ion in the spectrum of (I).

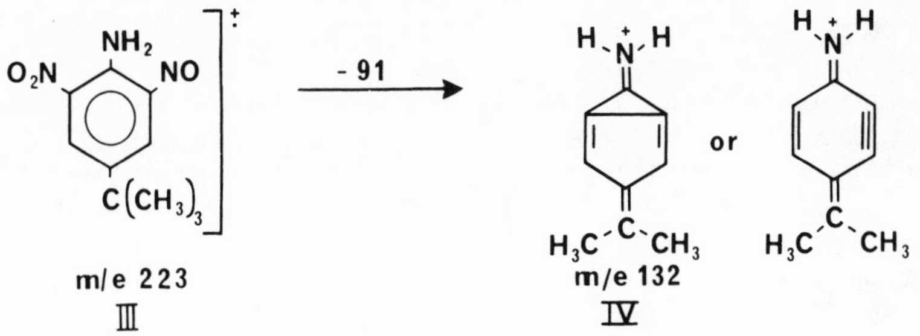

Figure 10

At m/e values less than 132 there are many fragments derived from the sequential expulsion of C and H atoms from the base peak.

N-Alkyl derivatives of 2,6-dinitroanilines show altered mass spectra through production of extremely abundant ions which result from α-cleavage of an alkyl carbon-carbon bond attached to the nitrogen atom. For example, the base peak of N-sec-butyl-4-tert-butyl-2,6-dinitroaniline at m/e 266 corresponds to the expulsion of a C_2H_5 radical from the secondary butyl group (Figure 11). Several herbicides of related structure possess this special feature. For example, trifluralin (a N-di-n-propylaniline deriv-ative) has M = 335, the base peak at m/e 306 [M-29] corresponds to loss of a C_2H_5 radical: there is also a strong ion at m/e 264, [M-71], which entails the loss of the fragments C_2H_5 and C_3H_6 and involves a rearrangement. N,N-Bis(2-chloroethyl)2,6-dinitro-p-toluidine, M = 321, has base peak at m/e 272 [M-CH₂Cl]. There is no peak of comparable intensity analogous to the [M-71] peak of trifluralin, indicating that the rearrangement pathway is probably influenced by the presence of the chlorine atom. This fragmen-

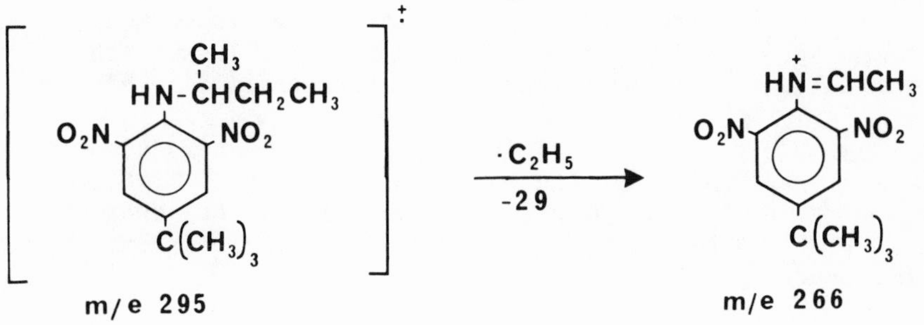

Figure 11

tation pattern is useful for recognizing modifications of the
N-alkyl moiety. For example, a microbial metabolite of N-sec-
butyl-4-tert-butyl-2,6-dinitroaniline had a molecular ion M = 311.
High-resolution mass spectrometry indicated the addition of an
oxygen atom to the parent molecule. The base peak was at m/e 266
(as in the parent compound) and its composition corresponded to
[M-C$_2$H$_5$O]. The metabolite was formed by hydroxylation of the
N-alkyl group at either the most remote or the penultimate carbon
atom (Kearney et al., 1972).

   Several metabolites and photoalteration products of N-alkyl-
2,6-dinitroanilines are formed by reaction between the N-alkyl
substituent and the nitro (or reduced nitro) group. The resultant
compounds may be benzimidazoles, benzimidazole N-oxides, or related
compounds. Their mass spectral fragmentation patterns frequently
indicate the presence of a stable aromatic system by a strong
molecular ion. Their spectra show some characteristic features.

   2-Methylbenzimidazole (V) (Figure 12) shows a strong peak at
the molecular ion M = 132 which is also the base peak. The next
most intense peak occurs at M-1, corresponding to loss of H. A
metastable peak at m/e 130 supports this pathway. Further losses
are indicated by peaks at m/e 118 [M-14], 104 [M-28], 90, 91 and 92
[M-42, 41 and 40].

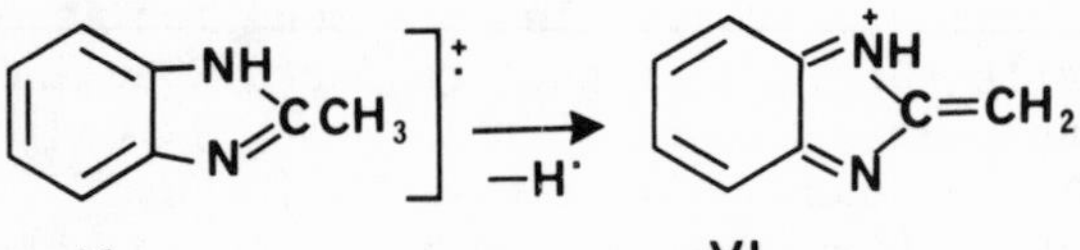

Figure 12

   The peak at M-1 shows that the fragmentation parallels that of
the 2-methylindoles (Beynon, 1960). A stable ion is formed (VI).
There is no apparent loss of CH$_3$ from 2-methylindoles and this
fragmentation path is also absent in 2-methylbenzimidazole.

   Benzimidazole itself (VII) (Figure 13), has a strong molecular
ion, M = 118, which is also the base peak. At m/e 91 is a peak
corresponding to the loss of HCN (27 mass units). At lower masses
there are peaks at m/e 63, 64, 69, 70 and 71. There is an extremely
small M-1 peak.

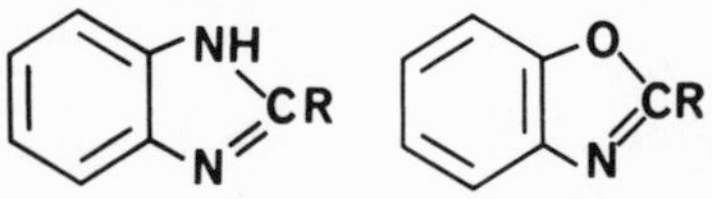

VI R = CH$_3$
VII R = H

VIII R = CH$_3$
IX R = H

Figure 13

The mass spectra of the oxygen analogue provide an interesting contrast. Elimination of oxygen containing fragments is preferred, since there is a tendency for the charge to remain localized on the nitrogen atom. 2-Methylbenzoxazole (VIII) has an intense molecular ion, M = 133 (the base peak), and a very small peak at M-1. Losses of 28 and 29 (CO and CHO) are significant and peaks at m/e 105 and 104 are observed. There is also a metastable ion at m/e 82.89 to substantiate the loss of CO. Other fragment ions are at m/e 92, 79, 63 and 64 and there are doubly-charged ions at m/e 66.5 and m/e 52.5 corresponding to the molecular ion m/e 133 and the ion at m/e 105 [M-28]. Loss of CH$_3$ is not observed. The spectrum of benzoxazole (IX) shows the strong molecular ion expected (M = 119). A peak at m/e 91 represents loss of CO and strong peaks at m/e 63 and 64 presumably correspond to C$_5$H$_3$ and C$_5$H$_4$ ions observed in this group of heterocyclic compounds.

It is interesting to apply these considerations to the mass spectrum of trifluralin metabolites. The major fragments in the mass spectrum of 2-ethyl-4-nitro-6-(trifluoromethyl)benzimidazole (X) (Figure 15) may be explained by reference to the spectrum of 2-ethylbenzimidazole. The M-1 peak of 2-ethylbenzimidazole is also the base peak and can be stabilized through an ion structure analogous to (VI). An alternative ion structure may be postulated for the M-1 fragment based on the fragmentation pathway of 2-ethylpyridine (Porter and Baldas, 1971). Hydrogen can be lost by γ-fission and the resultant ion analogous to (XI) (Figure 14) would be stabilized by cyclization. The spectrum of 2-ethyl-4-nitro-3-propyl-6-trifluorobenzimidazole (XII) (Figure 15), the N-propyl analogue of (X) shows a more complex fragmentation pattern. The molecular ion, although intense, is not the base peak, nor is there a loss of hydrogen from the molecular ion. The base peak is at m/e 213, [M-88], probably because of the loss of [C$_3$H$_6$+ NO$_2$], and there is a large peak at m/e 244, [M-57], which represents the loss of a propyl substituent together with a nitrogen atom from the imidazole ring. Below m/e 214, compounds (X) and (XII) may share a common pathway.

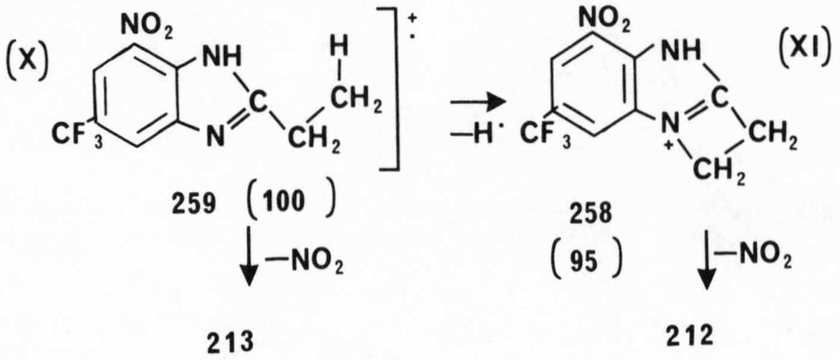

Figure 14

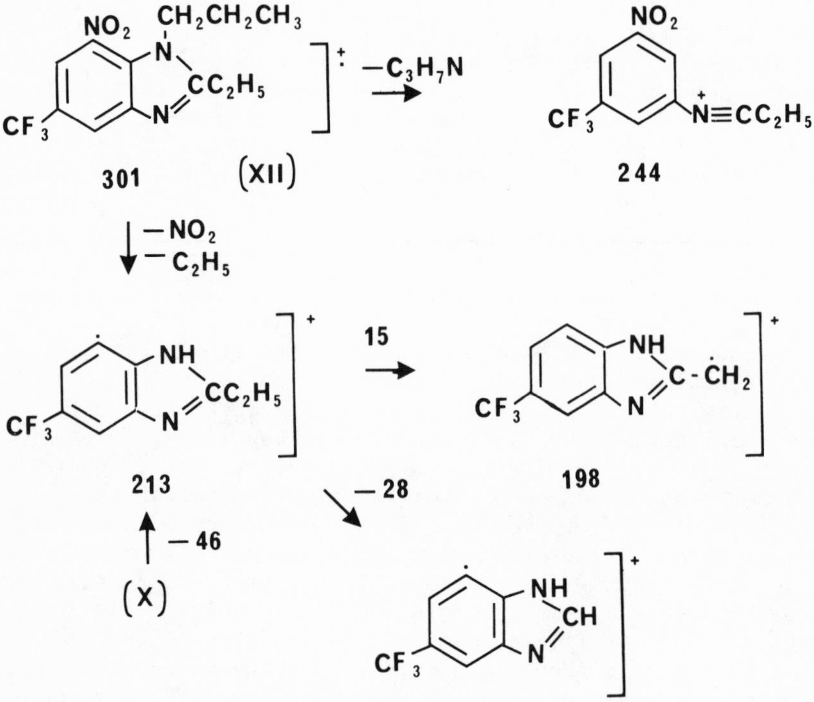

Figure 15

The foregoing discussion has provided a brief summary of some
of the important features of the spectra examined. In general the
formation of stable ions can most frequently be accounted for by
the location of a positive charge on the nitrogen atom. In addition,
the formation of an abundant molecular ion is a useful distinguishing
feature of some benzimidazole derivatives. Although the spectra are
generally complex, high resolution measurements permit rapid identification of several common fragmentation pathways.

LITERATURE CITED

1. Aloisi, G. G., E. Bordignon, and A. Signor, J. Chem. Soc.
    (Perkin Trans II) 2218, 1972.
2. Baldwin, M. A., A. G. Loudon, A. Maccoll, D. Smith, and A.
    Riberia, Chem. Commun., 351 (1967).
3. Barltrop, J. A., and N. J. Bunce, J. Chem. Soc. (C), 1968,
    1467.
4. Beynon, J. H., "Mass Spectrometry and its Application to
    Organic Chemistry", Elsevier, Amsterdam, 1960.
5. Budzikiewicz, H., C. Djerassi, and D. H. Williams, Mass
    Spectrometry of Organic Compounds, Holden-Day, Inc., San
    Francisco, 1967.
6. Bursey, M. M., Organic Mass Spectrometry, 2, 907, (1969).
7. Cable, J., S. A. Kagal, J. K. MacLeod, Organic Mass Spectrom-
    etry 6, 301 (1972).
8. Ciamician, G., and P. Silber, Atti. Accad. Lincei, 38, 3113
    (1905).
9. Crosby, D. G., and K. W. Moilanan, 163rd National Meeting
    American Chemical Society, Boston, Mass., April 11, 1972.
10. Davidson, R. S., S. K. Korkut, and P. R. Steiner, Chem. Commun.,
    1052, (1971).
11. Day, E. W., Unpublished, Cited by Probst, G. W. and Tepe, J. B.
    in Kearney, P. C., Kaufman, D. D., Eds. "Degradation of
    Herbicides", M. Dekker, New York, 1969.
12. Fielden, R., O. Meth-Cohn, and H. Suschitzky, Tetrahedron Lett.,
    1970, 1229.
13. Golab, T., Private Communication.
14. Kearney, P. C., and D. D. Kaufman (editors), Degradation of
    Herbicides, Marcel Dekker, Inc., New York, 1969.
15. Kearney, P. C., J. R. Plimmer, and V. P. Williams, 164th
    National Meeting American Chemical Society, New York, New York,
    August 29, 1972.
16. Laanio, T. L., P. C. Kearney, and D. D. Kaufman, 164th
    National Meeting American Chemical Society, New York, New
    York, August 29, 1972.

17. Leitis, E., and D. G. Crosby, American Chemical Society, Western Regional Meeting, San Francisco, California, October 1972.
18. McMahon, R. E., Tetrahedron Lett., 1966, 2307.
19. Meth-Cohn, O., Tetrahedron Lett., 1970, 1235.
20. Newsom, H. C., and W. G. Woods, J. Agr. Food Chem., in press.
21. Plimmer, J. R., and U. I. Klingebiel, 164th National Meeting, American Chemical Society, New York, New York, August 29, 1972.
22. Porter, Q. N. and J. Baldas, Mass Spectrometry of Heterocyclic Compounds, Wiley-Interscience, New York, 1971.
23. Preston, P. N., G. Tennant, Chem. Revs. 72, 672 (1972).
24. Russell, D. W., J. Chem. Soc., 1963, 894; 1964, 2829.
25. Schroll, G., R. G. Cooks, P. Klemmensen, and S. O. Lawesson, Arkiv Kemi, 28, 413 (1968).
26. Smith, R. A., 164th National Meeting American Chemical Society, New York, New York, August 29, 1972.
27. Trotter, W., and A. C. Testa, J. Amer. Chem. Soc., 90, 7044, (1968).

DETERMINATION OF METASTABLE TRANSITIONS IN THE MASS SPECTRA OF PESTICIDES BY DIRECT ANALYSIS OF DAUGHTER IONS

Francis J. Biros and James F. Ryan, III

Primate & Pesticides Effects Laboratory
U. S. Environmental Protection Agency
P.O. Box 490
Perrine, Florida 33157

INTRODUCTION

Observation and analysis of metastable transitions in the mass spectra of organic pesticidal compounds are important not only for studies concerned with the mechanism of fragmentation of ions formed in the mass spectrometer, but also provide information useful for the complete interpretation of mass spectra. Considerations of parent-product (daughter) ion relationships resulting from metastable studies are extremely valuable if the structure of the molecule must be reconstructed from the fragments. For example, a metastable peak relating two peaks in the fragmentation spectrum indicates that these two peaks arise from one and the same substance. Further, the smaller mass fragment of the pair must of necessity be composed entirely of atoms present in the larger mass fragment from which it arose by a unimolecular reaction mechanism involving generally, simple bond fission or electron transfer.

This approach may frequently be necessary in the determination of structure of unknown pesticide metabolites and characterization of pesticide environmental degradation products.

Metastable ions are formed in the mass spectrometer by decomposition of the parent species (which may or may not be the molecular ion) after complete acceleration from the ion source. In a standard Nier-Johnson design double focusing mass spectrometer, with the electrostatic sector (ESA) preceding the magnetic analyzer, decomposition of ions may occur either in the field-free region preceding the ESA (FF1), or immediately following it (FF2). With this configuration, studies are limited to those examining decompositions occurring in

113

the FFl region. These ions will undergo acceleration as the parent
ions, but mass analysis as product ions. Metastable ions of this
nature are collected by the electron multiplier at an apparent mass
of m*, i.e., M_2^2/M_1 where M_1 is the mass of the parent (precursor)
ion and M_2 is the mass of the product ion. The importance of
metastable ion relationships in confirming fragmentation patterns
and thus aiding molecular structure characterizations by mass spectrometry has resulted in the development of two techniques for
exclusively examining metastable ions by mass spectrometry.

By increasing the accelerating voltage to compensate for the
kinetic energy lost to the neutral fragment in the decomposition
$M_1^+ \longrightarrow M_2^+ + (M_1 - M_2)$, the product ion may be transmitted
through the electrostatic sector and collected by the beam monitor.
The entire metastable spectrum for any organic compound may be
obtained by appropriately varying the accelerating voltage. Frequently this will provide a fingerprint spectrum of the compound
of interest which is of value in characterization studies. This
method corresponds to the conventional defocusing technique which
has been used for some time.[1,2]

The second technique, initially described by Beynon et al.,[3,4,5]
requires scanning of the electrostatic analyzer at constant accelerating potential to permit product ions to be transmitted to an
electron multiplier which is strategically positioned following the
ESA. For each metastable decomposition, the ESA potential must be
decreased to accommodate the kinetic energy of the ion. The resultant ion kinetic energy (IKE) spectrum displays the kinetic
energy as a function of the abundance of each daughter ion produced
in the mass spectrum. The spectrum essentially summarizes all
metastable transitions in a single mass spectrum. IKE spectrometry
has been found to provide valuable information concerning the
kinetic and energetic characteristics of ions and ionic reactions
in the gas phase in addition to providing a unique fingerprint
spectrum in a manner analogous to conventional metastable spectra
obtained by accelerating potential scanning. Recent reports have
described the technique and utility of IKE spectrometry in the
characterization of physical and chemical properties of organic
compounds. Of particular interest and application to pesticide
chemistry is the potential analytical utility of IKE spectrometry
in providing an unequivocal determination of residues of toxic
substances and in distinguishing isomeric compounds.[6] This latter
application is of special importance as conventional electron impact
(EI) spectrometry[7] and in some cases positive and negative chemical
ionization mass spectrometry[8,9,10] provide only equivocal results
in the characterization of isomeric pairs of pesticidal compounds.
Other applications include determination of isotopic incorporation[11],
studies in multiple charged ions[12], isotope effects, charge-exchange
reactions and the measurement of dynamic physical parameters of
atoms and molecules.[13]

A third technique for the study of metastable spectra is
possible with mass spectrometers which have reversed field arrangements, i.e., an electrostatic analyzer following the magnetic
sector (Fig. 1). The magnetic field is adjusted for either molecular
ions or fragment ions of any m/e value and if these ions are subject
to further decomposition in the second field free region (FF2), all
such ions may be recorded by decreasing the ESA voltage in a single
scan. The resulting spectrum has been termed both a direct analysis
of daughter ion spectrum (DADI)[14], and a mass analyzed ion kinetic
energy spectrum (MIKES)[5]; with ESA scanning, the mass scale is
given by $M_2 = M_1 \cdot E/E_0$. The several advantages of the DADI technique include the fact that all ion species enter the field free
region in front of the ESA with maximum energy, the sensitivity is
high and consistent throughout the experiment, there is no possibility of interference between different transitions, and there is
no limit to the ratio M_1/M_2 that can be studied. In fact, all
product ions formed by unimolecular decomposition of the selected
precursor ion are recorded. The effective result is a mass analyzed
product ion spectrum (MAPS) for any peak in a typical mass spectrum.
The present investigation describes the application of the DADI-MAPS
technique to the analysis and characterization of a group of structurally diverse pesticidal compounds. On the basis of these results,
it is proposed that DADI-MAPS may (a) provide a unique complementary
technique for analysis of pesticide molecules, (b) provide characterization of molecular structure which may be of value in determination

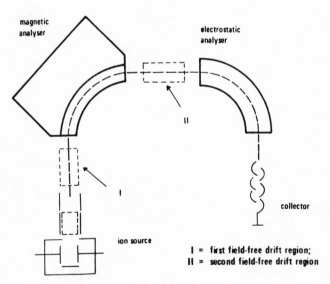

Figure 1.   Block diagram of double focusing mass spectrometer with
            reversed field arrangement illustrating field free re-
            gions (FF1 and FF2) where ionic reactions may be exam-
            ined.

of structure of closely related molecules produced by human and animal metabolism or environmental degradation, and (c) offer a means for distinguishing isomeric pairs of pesticides and other toxic substances.

EXPERIMENTAL

The pesticidal materials examined in this study were high purity analytical grade crystalline standards used as received from the manufacturer or recrystallized as indicated by gas chromatographic purity checks. The individual pesticides included in the study are as follows:

Abate: O,O,O',O'-Tetramethyl O,O'-thiodi-p-phenylene phosphorothioate

Barban: 4-Chloro-2-butynyl-m-chlorocarbanilate

Benomyl: Methyl-1-(butyl carbamoyl) 2-benzimidazole carbamate

Dicapthon: O,O'-Dimethyl O-(2-chloro-4-nitrophenyl)phosphorothioate

Diphenamid: N,N-Dimethyl 2,2-diphenylacetamide

Mass spectral data were obtained with a Varian CH-5 double focusing, reversed field mass spectrometer equipped with electron impact ion source and automatic ESA voltage scanner for observation of metastable peaks produced in the second field free region of the mass spectrometer. Samples were introduced directly into the ionization chamber with a temperature controlled probe. Sufficient heat was applied to maintain an adequate sample vapor pressure in the ion source during the determination of the DADI (MAPS) spectra. Low resolution spectra were obtained first, then the major ions in the spectrum were selected for study, the magnetic analyzer tuned to the appropriate mass spectral peak, and the ESA scanned from 500 V, nominal, to lower values. The voltage measurements were made employing a Hewlett Packard #3490A digital multimeter. The resulting DADI spectrum was essentially a plot of kinetic energy versus abundance of product ions produced in region FF2 from the selected major spectrum peak. The mass of each observed product ion was calculated from the relation $M_2 = M_1 \cdot E/E_0$. Generally, 8-10 individual DADI spectra were obtained for each compound. The detailed results of these experiments are tabulated in the Appendix. Characterization of the fragments exhibited in the DADI spectra was facilitated by high resolution measurements of the same peaks in the electron impact spectra.

RESULTS & DISCUSSION

This study included examination of two organophosphorus compounds, a carbamate, carbanilate and an aromatic dimethyl acetamide.
Each compound was selected to represent a typical member of a specific
class of pesticidal compounds which frequently provide complex EI
fragmentation spectra. The normally focused sample ion beam is
observed at 3kV accelerating voltage and 500 V, nominal, electrostatic sector voltage. The peak selected is located by increasing
the magnetic field to the desired m/e value. An intense ion signal
is thus produced at the left of the DADI spectrum, which is recorded
by a potentiometric recorder. Scanning to lower ESA voltage values,
product ion peaks are produced representing decompositions of the
selected ion in the FF2 region of the mass spectrometer. Each peak
represents the loss of a neutral fragment from the precursor ion.
Generally, most product ions observed in this fashion are found in
the major EI spectrum. Assignment of structures of the product ions
(as well as precursor ions) in each instance was verified by high
resolution measurements made on the major EI spectrum.

The EI mass spectrum of Abate (Fig. 2) exhibited an intense
molecular ion of m/e 466 with fragments observed at m/e 372, 357,
339, 325, 231, and 203. The remaining peaks in the spectrum correspond to previously observed decompositions of $CH_3OP(O)SCH_3^+$ and
like species. DADI spectral data for peaks m/e 466, the molecular
ion, and m/e 372, a rearrangement ion produced by elimination of
CH_3OPO_2 from the molecular ion,are illustrated in Fig. 3. Major
product ions for m/e 466 as observed in the spectrum include m/e
433, formed by loss of SH, and m/e 372, an ion which must be produced by elimination of CH_3OPO_2 frcm the phosphorothiolate parent
rearrangement ion[15] followed by methyl migration.[16] This rearrangement has been postulated to occur by way of a five membered ring
heterocyclic intermediate.[17] The exact counterpart ion produced
by methyl migration from the unrearranged molecular ion is found
at m/e 356. The remaining product ions are formed by straightforward bond fission involving loss of sulfur (m/e 339), loss of
a dimethyl phosphorothioate group (m/e 325), and cleavage of the
phenyl thioether bond (m/e 203) as illustrated in Fig. 4.

The carbanilate class of pesticides is represented by the
compound barban. The low resolution mass spectrum is reproduced
in Fig. 5. Major ions observed included the molecular ion m/e 257
(20%), and fragment ions m/e 222 (85%), m/e 153 (25%), and m/e 143
(40%). Spectra produced by direct analysis of daughter ions with
the magnetic sector focused on m/e 257 and m/e 222 are illustrated
in Fig. 6. Each ion may be observed to be precursor for three
product ions. Decomposition of m/e 257, the molecular ion,re-
sulted in formation of peaks at m/e 222, by loss of Cl, m/e 153,
m-chlorophenyl isocyanate ion, by elimination of 4-chloro-2-butynyl

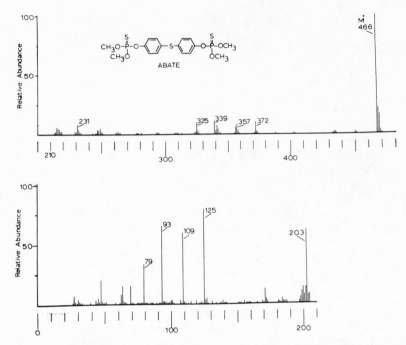

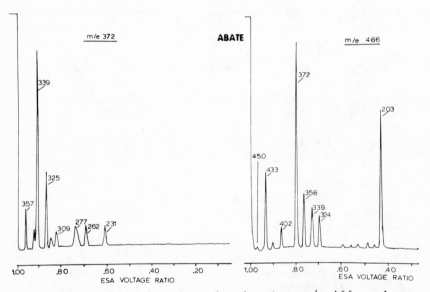

Figure 2.   Low resolution electron impact mass spectrum of the
organophosphate pesticide Abate.

Figure 3.   DADI spectra for the molecular ion, m/e 466, and a
major rearrangement ion, m/e 372, in the mass spectrum
of Abate.

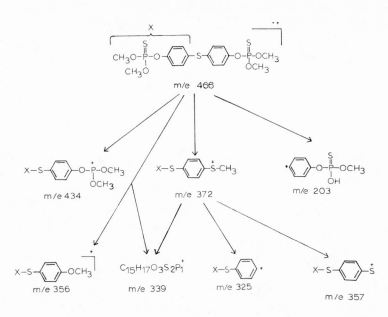

Figure 4. Proposed major fragmentation processes for the molecular ion, m/e 466, in the electron impact mass spectrum of Abate.

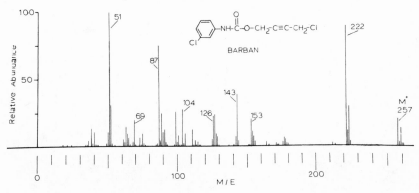

Figure 5. Low resolution electron impact mass spectrum of the carbanilate pesticide barban.

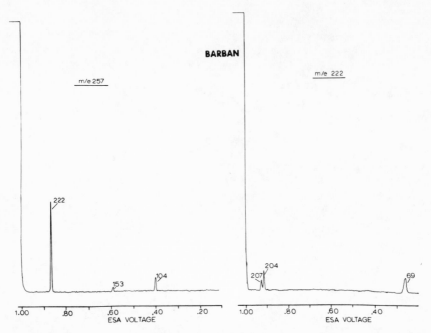

Figure 6.   DADI spectra for the molecular ion, m/e 257, and the (M–Cl)+ ion, m/e 222, in the mass spectrum of barban.

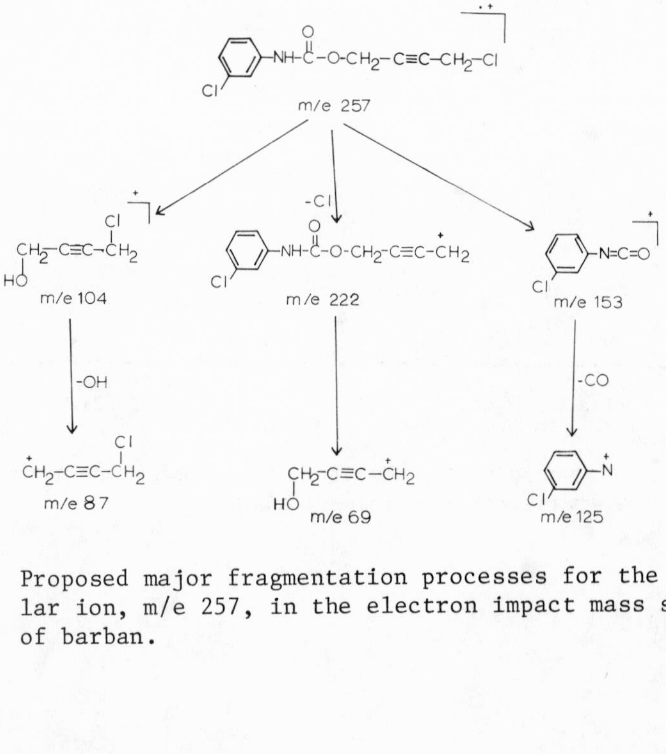

Figure 7.   Proposed major fragmentation processes for the molecular ion, m/e 257, in the electron impact mass spectrum of barban.

alcohol, and m/e 104, the complementary 4-chloro-2-butynyl alcohol
cation (Fig. 7). Product ions resulting directly from the decomposition of m/e 222 include m/e 207 formed by loss of a methyl radical,
m/e 204, formed by loss of water, and m/e 69, cleavage at the carbonyl
carbon, followed by hydrogen rearrangement to yield the butynyl
alcohol cation.

The low resolution mass spectrum of benomyl is reproduced in
Fig. 8. No molecular ion was observed at m/e 290. Significant
fragment ions, however, were found at m/e 191, 159, 146, 132, and
105. The DADI spectra obtained for fragments m/e 191 and m/e 169
are illustrated in Fig. 9. Decomposition of m/e 191, by loss of
CO_2CH_3, CH_3OH, and HCO_2, the latter process accompanied by migration
of the methyl group to the 1-imidazole nitrogen, results in the
formation of fragment ions of m/e 159, m/e 146, and m/e 133. The
m/e 159 precursor ion was observed (Fig. 10) to decompose into two
species at m/e 147 and m/e 131; the latter ion is formed by loss of
CO, as verified by high resolution measurements.

Dicapthon represented the second organophosphate examined in
this study. A molecular ion (m/e 297) was not observed in the
low resolution mass spectrum (Fig. 11). The mass analyzed ion
kinetic energy spectra observed for fragment ions m/e 262 and m/e
125 are illustrated in Fig. 12. The fragment ion of m/e 262 is
formed by facile loss of Cl from the molecule ion. Reference to
Fig. 13 provides the major specific molecular decomposition pathways adduced for this compound based on the DADI spectra and high
resolution accurate mass measurements. Product ions resulting
directly from decomposition of the m/e 262 ion include m/e 247,
formed by loss of methyl radical, m/e 232, formed by loss of nitric
oxide, NO, radical, and m/e 216 formed by loss of nitrogen dioxide,
NO_2. Decomposition pathways for the dimethyl phosphorothioate ion,
m/e 125, may also be determined. From the data reproduced in Fig. 12,
it is apparent that complete equilibration of $CH_3OP(S)$ OCH_3 to
$CH_3SP(O)$ OCH_3 probably occurs. Major product ions include: m/e 47,
SCH_3+; m/e 62, CH_3OP+; m/e 79, CH_3OPOH+; and m/e 95, CH_3OPSH+.

The low resolution mass spectrum of diphenamid illustrated
in Fig. 14, exhibited a molecular ion at m/e 239, and fragment peaks
at m/e 167, m/e 165, m/e 152, and a base peak of m/e 72. Ions
selected for DADI analysis and their resultant spectra are presented
in Fig. 15, and include m/e 239 and m/e 167. Decomposition of m/e
239 occurs by loss of methyl radical to produce a fragment of m/e
224, loss of N,N-dimethylisocyanate radical to produce an ion of
m/e 167, the diphenylmethyl cation, and finally, the base peak con-
jugate ion for this process of m/e 72, $(CH_3)_2$ $NCO+$. The fragment
ion of m/e 167 is precursor to the fluorenyl cation of m/e 165[18],
and also the substituted cyclobutadienyl cation at m/e 152, possibly

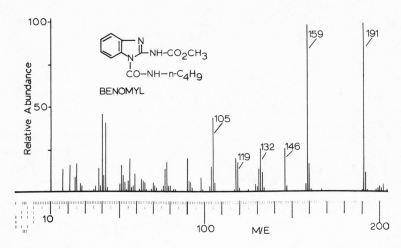

Figure 8.   Low resolution electron impact mass spectrum of the
            carbamate pesticide benomyl.

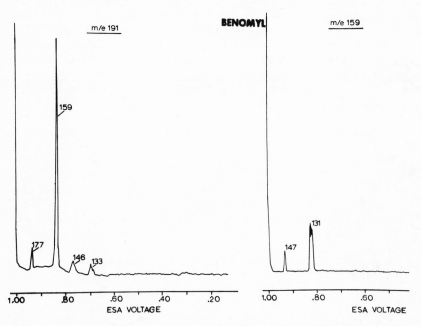

Figure 9.   DADI spectra for the $(M-C_5H_9NO)^+$ ion, m/e 191, and the
            $(191-CH_4O)^+$ ion, m/e 159, in the mass spectrum of
            benomyl.

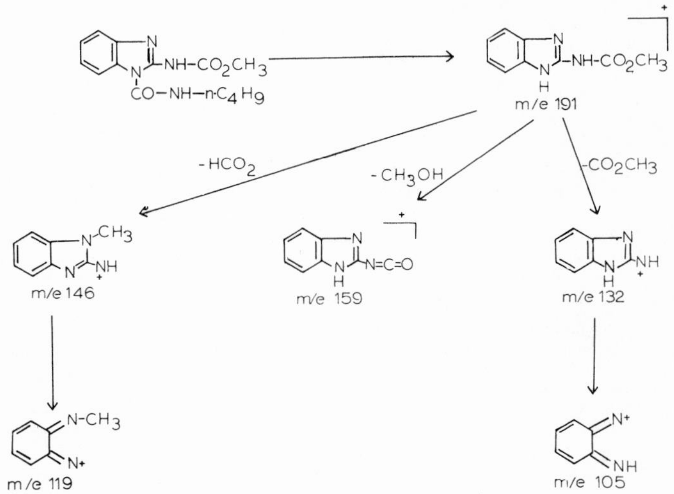

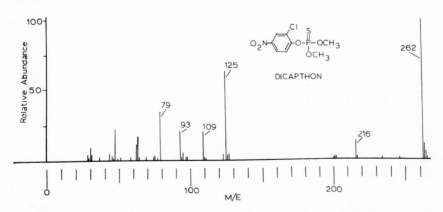

Figure 10. Proposed major fragmentation processes for the $(M-C_5H_9NO)^+$ ion, m/e 191, in the mass spectrum of benomyl.

Figure 11. Low resolution electron impact mass spectrum of the organophosphate pesticide dicapthon.

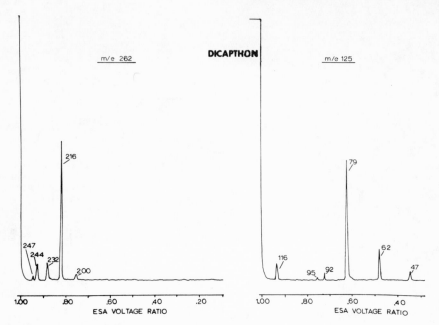

Figure 12. DADI spectra for the (M–Cl)+ ion, m/e 262, and the
$C_2H_6O_2PS+$ ion, m/e 125, in the mass spectrum of dicapthon.

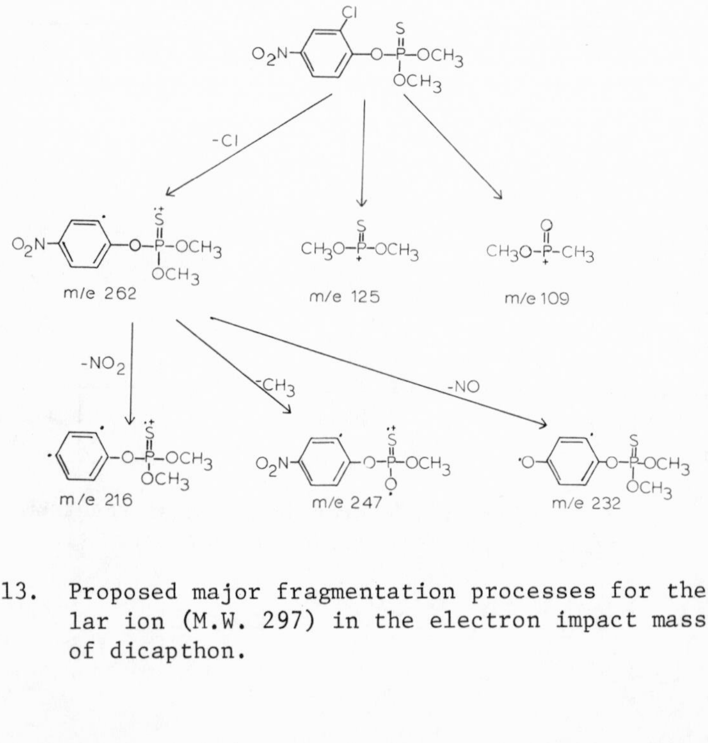

Figure 13. Proposed major fragmentation processes for the molecular ion (M.W. 297) in the electron impact mass spectrum
of dicapthon.

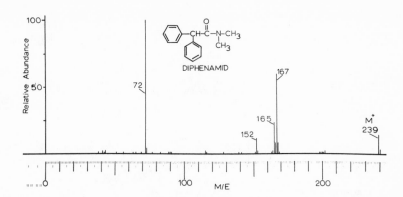

Figure 14.   Low resolution electron impact mass spectrum of the dimethylacetamide pesticide diphenamid.

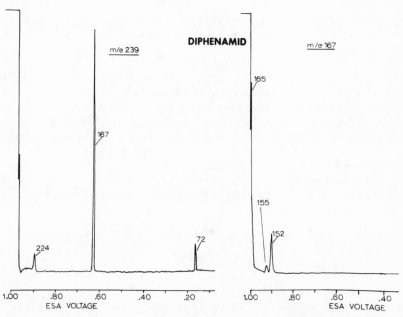

Figure 15.   DADI spectra for the molecular ion, m/e 239, and the $(M-C_3H_6NO)+$ ion, m/e 167, in the mass spectrum of diphenamid.

via a phenyltropylium intermediate as proposed by other investigators[19], (Fig. 16).

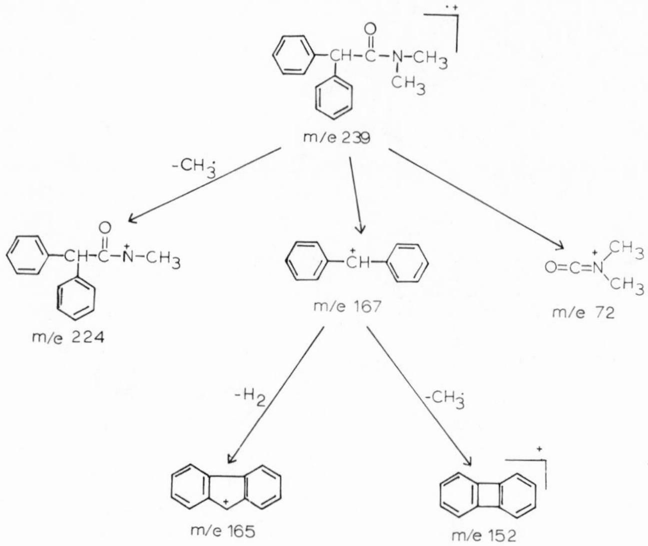

Figure 16.   Proposed major fragmentation processes for the molecular ion, m/e 239, in the electron impact mass spectrum of diphenamid.

     In summary, the DADI technique provides an extremely valuable adjunct to the study of fragmentation mechanisms in the EI impact induced mass spectra of pesticidal compounds. This approach should prove invaluable in correlating mass spectral behavior and molecular structure of pesticide metabolites and degradation products of unknown structure derived from the parent materials examined in this and related studies. Potential analytical applications are presently being investigated and particularly with respect to the ability of the DADI technique to distinguish isomeric pesticidal compounds.

REFERENCES

1.   M. Barber and R. M. Elliot, Comparison of Metastable Spectra from Single and Double-Focusing Mass Spectrometers, Abstracts, ASTM Committee E-14, 12th Annual Conference on Mass Spectrometry and Allied Topics, Montreal, Canada, June 1964.

2.   J. H. Beynon, R. A. Saunders, and A. E. Williams, Nature, 204, 67 (1964).

3. J. H. Beynon, J. W. Amy, and W. E. Baitinger, Chem. Commun.,
    723 (1969).

4. A. H. Struck and H. W. Major, Enhanced Metastable Ion
    Measurement by Defocused Operation of Double Focusing Mass
    Spectrometers, Abstracts, ASTM Committee E-14, 17th Annual
    Conference on Mass Spectrometry and Allied Topics, Dallas,
    Texas, May 1969.

5. J. H. Beynon and R. G. Cooks, Res./Develop., 22, 26 (1971).

6. S. Safe, O. Hutzinger, and W. D. Jamieson, The Ion Kinetic
    Energy Spectra of Pesticides and Related Compounds, Abstracts,
    165th National Meeting of the American Chemical Society, Dallas,
    Texas, April 1973, PEST-57.

7. F. J. Biros, Residue Reviews, 40, 1 (1971).

8. F. J. Biros, R. C. Dougherty, and J. Dalton, Org. Mass Spectrom.,
    6, 1161 (1972).

9. R. C. Dougherty, J. Dalton, and F. J. Biros, Org. Mass Spectrom.,
    6, 1171 (1972).

10. R. C. Dougherty, J. D. Roberts, and F. J. Biros, Positive and
    Negative Chemical Ionization Mass Spectra of Polychlorinated
    Pesticides, Abstracts, 165th National Meeting of the American
    Chemical Society, Dallas, Texas, April 1973, PEST-59.

11. J. H. Beynon, J. E. Corn, W. E. Baitinger, R. M. Caprioli, and
    R. A. Benkeser, Org. Mass Spectrom., 3, 1371 (1970).

12. J. H. Beynon, R. M. Caprioli, and J. W. Richardson, J. Amer.
    Chem. Soc., 93, 1852 (1971).

13. R. H. Shapiro, K. B. Tomer, R. M. Caprioli, and J. H. Beynon,
    Org. Mass Spectrom., 3, 1333 (1970).

14. K. H. Maurer, C. Brunnee, G. Kappus, K. Habfast, U. Schroder,
    and P. Schulze, Direct Analysis of Daughter Ions Arising from
    Metastable Decompositions, Abstracts, 19th Annual Conference
    of the American Society for Mass Spectrometry, Atlanta, Georgia,
    May 1971, p. K9.

15. F. J. Biros and R. T. Ross, Fragmentation Processes in the Mass
    Spectra of Trialkyl Phosphates, Phosphorothionates, Phosphoro-
    thiolates, and Phosphorodithioates, Abstracts, 18th Annual
    Conference of the American Society for Mass Spectrometry, San
    Francisco, Calif., June 1970, p. G3.

16. J. N. Damico, J. Ass. Offic. Anal. Chem., 49, 1027 (1966).

17. W. A. Laurie, M. E. Wacks, R. W. Anderson, J. F. Dietz, and B.
     Lacy, Mass Spectral Rearrangements in Organophosphorus Pesticides,
     Abstracts, 18th Annual Conference of the American Society for
     Mass Spectrometry, San Francisco, Calif., June 1970, p. B1.

18. J. Jörg, R. Houriet, and G. Spiteller, Monatsh. Chem., 97, 1064
     (1966).

19. K. D. Berlin and R. D. Shupe, Org. Mass Spectrom., 2, 447 (1969).

APPENDIX

TABLE 1

Decompositions in the Second Field Free Region of the Mass
Spectrometer of Selected Ions in the Mass Spectrum of Abate
(M.W. 466) Determined by DADI

Decomposition Process			Precursor Ion		Daughter Ion	Rel. Int. X 10^4
$C_{16}H_{20}O_6P_2S_3^+$	$\rightarrow$	$[C_{16}H_{19}O_6P_2S_3]^{+a}$	466	$\rightarrow$	465	416.67^b
	$\rightarrow$	$C_{16}H_{20}O_6P_2S_2^+$		$\rightarrow$	434	1.25
	$\rightarrow$	$[C_{16}H_{20}O_6P_2S]^+$		$\rightarrow$	402	0.54
	$\rightarrow$	$C_{15}H_{17}O_3PS_3^+$		$\rightarrow$	372	2.68
	$\rightarrow$	$C_{15}H_{17}O_4PS_2^+$		$\rightarrow$	356	0.95
	$\rightarrow$	$C_{15}H_{16}O_3PS_2^+$		$\rightarrow$	339	0.83
	$\rightarrow$	$C_{14}H_{14}O_3PS_2^+$		$\rightarrow$	325	0.60
	$\rightarrow$	$C_7H_8O_3PS^+$		$\rightarrow$	203	1.73
$C_{15}H_{17}O_3PS_3^+$	$\rightarrow$	$C_{15}H_{16}O_3PS_3^+$	372	$\rightarrow$	371	2.50
	$\rightarrow$	$C_{14}H_{14}O_3PS_3^+$		$\rightarrow$	357	2.50
	$\rightarrow$	$C_{15}H_{16}O_3PS_2^+$		$\rightarrow$	339	1.20
	$\rightarrow$	$C_{14}H_{14}O_3PS_2^+$		$\rightarrow$	325	7.50
	$\rightarrow$	$- -$		$\rightarrow$	308	1.50
	$\rightarrow$	$- -$		$\rightarrow$	277	0.25
	$\rightarrow$	$[C_9H_{11}O_3PS_2]^+$		$\rightarrow$	262	0.25
	$\rightarrow$	$C_{13}H_{11}S_2^+$		$\rightarrow$	231	2.00
$C_{14}H_{14}O_3PS_3^+$	$\rightarrow$	$[C_{14}H_{13}O_3PS_3]^+$	357	$\rightarrow$	355	2.27
	$\rightarrow$	$- -$		$\rightarrow$	342	4.54
	$\rightarrow$			$\rightarrow$	329	1.59
	$\rightarrow$	$C_{14}H_{14}O_3PS_2^+$		$\rightarrow$	325	4.30
	$\rightarrow$	$- -$		$\rightarrow$	293	1.50
	$\rightarrow$	$C_9H_{12}O_3PS_2^+$		$\rightarrow$	263	0.68
	$\rightarrow$	$C_{12}H_8S_3^+$		$\rightarrow$	248	1.44
	$\rightarrow$	$C_{12}H_8S_2^+$		$\rightarrow$	216	1.14
$C_{15}H_{17}O_4PS_2^+$	$\rightarrow$	$C_{15}H_{16}O_4PS_2^+$	356	$\rightarrow$	355	11.76
	$\rightarrow$	$C_{15}H_{15}O_4PS_2^+$		$\rightarrow$	354	1.50
	$\rightarrow$	$C_{14}H_{14}O_4PS_2^+$		$\rightarrow$	341	1.50
	$\rightarrow$	$C_{14}H_{13}O_3PS_2^+$		$\rightarrow$	324	3.00
	$\rightarrow$	$C_9H_{11}O_3PS_2^+$		$\rightarrow$	262	1.50
$C_{14}H_{14}O_4PS_2$	$\rightarrow$	$[C_{14}H_{13}O_4PS_2]^+$	341	$\rightarrow$	340	6.25
	$\rightarrow$	$- -$		$\rightarrow$	327	4.17
	$\rightarrow$	$C_{13}H_{11}S_2^+$		$\rightarrow$	231	7.91
	$\rightarrow$	$C_{12}H_8S_2^+$		$\rightarrow$	216	5.63
	$\rightarrow$	$C_9H_{12}OS_2^+$		$\rightarrow$	200	1.46

TABLE 1 (Cont.)

Decomposition Process		Precursor Ion		Daughter Ion	Rel. Int. X 10^4
$C_{15}H_{16}O_3PS_2+$	$\rightarrow$ $[C_{15}H_{15}O_3PS_2]+$	339	$\rightarrow$	338	9.38
	$\rightarrow$ $[C_{15}H_{14}O_3PS_2]+$		$\rightarrow$	337	6.25
	$\rightarrow$ $[C_{15}H_{13}O_3PS_2]+$		$\rightarrow$	336	4.10
	$\rightarrow$ $[C_{15}H_{16}O_2PS_2]+$		$\rightarrow$	323	1.46
	$\rightarrow$ $[C_{15}H_{16}O_3PS]+$		$\rightarrow$	307	1.46
	$\rightarrow$ $C_{13}H_{11}S_2+$		$\rightarrow$	231	1.46
$C_{13}H_{11}S_2+$	$\rightarrow$ $C_{13}H_{10}S_2+$	231	$\rightarrow$	230	7.50
	$\rightarrow$ $C_{13}H_9S_2+$		$\rightarrow$	229	2.50
	$\rightarrow$ $C_{12}H_8S_2+$		$\rightarrow$	216	2.25
	$\rightarrow$ --		$\rightarrow$	204	2.50
	$\rightarrow$ --		$\rightarrow$	187	3.25
$C_7H_8O_3PS+$	$\rightarrow$ $C_7H_7O_3PS+$	203	$\rightarrow$	202	1.50
	$\rightarrow$ --		$\rightarrow$	189	1.25
	$\rightarrow$ --		$\rightarrow$	173	1.75
	$\rightarrow$ $C_2H_6O_3P+$		$\rightarrow$	109	1.33
$C_7H_8O_3P+$	$\rightarrow$ $[C_7H_7O_3P]+$	171	$\rightarrow$	170	14.12
	$\rightarrow$ $[C_7H_6O_3P]+$		$\rightarrow$	169	4.12
	$\rightarrow$ $C_4H_4O_2PS+$		$\rightarrow$	147	5.88
	$\rightarrow$ --		$\rightarrow$	139	4.12
	$\rightarrow$ --		$\rightarrow$	127	5.29
	$\rightarrow$ --		$\rightarrow$	119	7.65
$C_2H_6O_2PS+$	$\rightarrow$ $[C_2H_5O_2PS]+$	125	$\rightarrow$	124	0.91
	$\rightarrow$ $[C_2H_4O_2PS]+$		$\rightarrow$	123	0.53
	$\rightarrow$ --		$\rightarrow$	116	0.76
	$\rightarrow$ --		$\rightarrow$	103	0.58
	$\rightarrow$ CH_4PS+		$\rightarrow$	79	1.27
	$\rightarrow$ CH_3OP+		$\rightarrow$	62	0.53
	$\rightarrow$ $OP+$		$\rightarrow$	47	0.36
C_2H_6OPS+	$\rightarrow$ --	109	$\rightarrow$	101	0.64
	$\rightarrow$ CH_4PS+		$\rightarrow$	79	1.25
$C_2H_6O_2P+$	$\rightarrow$ --	93	$\rightarrow$	87	1.87
	$\rightarrow$ CH_4OP+		$\rightarrow$	63	1.30

[a]Bracketed empirical composition data are not supported by high resolution mass measurements.

[b]Based on value of 100 for the precursor ion.

TABLE 2

Decompositions in the Second Field Free Region of the Mass
Spectrometer of Selected Ions in the Mass Spectrum of Barban
(M.W. 257) Determined by DADI

Decomposition Process		Precursor Ion	Daughter Ion	Rel. Int. X 10^4
$C_{11}H_9Cl_2NO_2+$	→ $C_{11}H_9ClNO_2+$	257	→ 222	12.50^b
	→ C_7H_4ClNO+		→ 153	1.06
	→ C_4H_5ClO+		→ 104	3.53
$C_{11}H_9ClNO_2+$	→ $[C_{11}H_8ClNO_2]+^a$	222	→ 221	0.60
	→ $[C_{10}H_6ClNO_2]+$		→ 207	0.12
	→ $[C_{11}H_9ClNO]+$		→ 204	0.60
	→ C_4H_5O+		→ 69	0.60
C_7H_5ClNO+	→ C_6H_5ClN+	154	→ 126	1.50
C_7H_4ClNO+	→ C_6H_4ClN+	153	→ 125	2.67
C_6H_6ClN+	→ C_6H_5ClN+	127	→ 126	1.00
	→ C_5H_4Cl+		→ 99	2.00
	→ C_6H_6N+		→ 92	0.67
C_6H_5ClN+	→ - -	126	→ 117	0.56
	→ C_5H_4Cl+		→ 99	2.22
	→ C_6H_5Cl+		→ 91	0.83
	→ C_6H_4N+		→ 90	1.67
C_4H_5ClO+	→ C_4H_4Cl+	104	→ 87	2.50
C_4H_4Cl+	→ C_4H_3+	87	→ 51	5.00

[a] Bracketed empirical composition data are not supported by high
resolution mass measurements.

[b] Based on a value of 100 for the precursor ion.

TABLE 3

Decompositions in the Second Field Free Region of the Mass
Spectrometer of Selected Ions in the Mass Spectrum of Benomyl
(M.W. 290) Determined by DADI

Decomposition Process			Precursor Ion		Daughter Ion	Rel. Int. X 10^4
$C_9H_9N_3O_2^+$	$\rightarrow$	$[C_8H_7N_3O_2]^{+a}$	191	$\rightarrow$	177	0.43^b
	$\rightarrow$	$C_8H_5N_3O^+$		$\rightarrow$	159	5.14
	$\rightarrow$	$C_8H_8N_3^+$		$\rightarrow$	146	0.29
	$\rightarrow$	$C_7H_7N_3^+$		$\rightarrow$	133	0.14
$C_8H_5N_3O^+$	$\rightarrow$	$- -$	159	$\rightarrow$	147	0.19
	$\rightarrow$	$C_7H_6N_3^+$		$\rightarrow$	131	0.56
$C_8H_8N_3^+$	$\rightarrow$	$C_8H_7N_3^+$	146	$\rightarrow$	145	3.81
	$\rightarrow$	$- -$		$\rightarrow$	135	0.48
	$\rightarrow$	$[C_8H_5N_2]^+$		$\rightarrow$	129	2.38
	$\rightarrow$	$C_7H_7N_2^+$		$\rightarrow$	119 ⎱	0.98
	$\rightarrow$	$C_7H_6N_2^+$		$\rightarrow$	118 ⎰	
$C_7H_7N_3^+$	$\rightarrow$	$- -$	132	$\rightarrow$	109	21.43
	$\rightarrow$	$C_6H_5N_2^+$		$\rightarrow$	105	7.14
$C_7H_6N_3^+$	$\rightarrow$	$- -$	131	$\rightarrow$	122	0.42
	$\rightarrow$	$- -$		$\rightarrow$	108	0.83
	$\rightarrow$	$C_6H_4N_2^+$		$\rightarrow$	104	2.92
$C_7H_7N_2^+$	$\rightarrow$	$C_6H_6N^+$	119	$\rightarrow$	92	2.86
$C_7H_6N_2^+$	$\rightarrow$	$C_7H_5N_2^+$	118	$\rightarrow$	117	1.67
	$\rightarrow$	$C_6H_5N^+$		$\rightarrow$	91 ⎱	3.80
	$\rightarrow$	$C_6H_4N^+$		$\rightarrow$	90 ⎰	
$C_6H_5N_2^+$	$\rightarrow$	$C_6H_4N_2^+$	105	$\rightarrow$	104	0.46
	$\rightarrow$	$C_5H_4N^+$		$\rightarrow$	78	0.92
$C_6H_4N^+$	$\rightarrow$	$C_6H_3N^+$	90	$\rightarrow$	89	0.21
	$\rightarrow$	$C_6H_2N^+$		$\rightarrow$	88	0.21
	$\rightarrow$	$C_5H_3^+$		$\rightarrow$	63	1.67
$C_5H_4N^+$	$\rightarrow$	$C_5H_3N^+$	78	$\rightarrow$	77	12.50
	$\rightarrow$	$C_4H_3^+$		$\rightarrow$	51	31.25
$C_4H_3^+$	$\rightarrow$	$C_4H_2^+$	51	$\rightarrow$	50	2.86

[a] Bracketed empirical composition data are not supported by high
resolution mass measurements.

[b] Based on a value of 100 for the precursor ion.

TABLE 4

Decompositions in the Second Field Free Region of the Mass
Spectrometer of Selected Ions in the Mass Spectrum of Dicapthon
(M.W. 297) Determined by DADI

Decomposition Process			Precursor Ion (m/e)		Daughter Ion (m/e)	Rel. Int. X 10^4
$C_8H_9NO_5PS+$	→	$[C_7H_6NO_5PS]+^a$	262	→	247	0.27^b
	→	$[C_8H_7NO_4PS]+$		→	244	0.54
	→	$C_8H_9O_4PS+$		→	232	0.45
	→	$C_8H_9O_3PS+$		→	216	2.70
	→	$C_7H_7NO_2PS+$		→	200	0.27
$C_8H_9O_3PS+$	→	$C_8H_8O_3PS+$	216	→	215	3.33
	→	$[C_7H_6O_3PS]+$		→	201	5.00
	→	− −		→	188	2.50
$C_2H_6O_2PS+$	→	− −	125	→	116	0.98
	→	CH_4O_2P+		→	79	2.06
	→	CH_3OP+		→	62	1.13
	→	$PO+$		→	47	0.11
$C_2H_6O_3P+$	→	CH_4O_2P+	109	→	79	1.30
$C_2H_6O_2P+$	→	CH_4OP+	93	→	63	5.00
CH_4O_2P+	→	− −	79	→	67	0.22

---

[a] Bracketed empirical composition data are not supported by high
resolution mass measurements.

[b] Based on a value of 100 for the precursor ion.

TABLE 5

Decompositions in the Second Field Free Region of the Mass
Spectrometer of Selected Ions in the Mass Spectrum of
Diphenamid (M.W. 239) Determined by DADI

Decomposition Process		Precursor Ion		Daughter Ion	Rel. Int. X 10^4
$C_{16}H_{17}NO^+$	$\rightarrow$ $C_{16}H_{16}NO^+$	239	$\rightarrow$	238	2.11^b
	$\rightarrow$ $[C_{16}H_{17}N]^{+a}$		$\rightarrow$	223	1.41
	$\rightarrow$ $C_{13}H_{11}^+$		$\rightarrow$	167	4.74
	$\rightarrow$ $C_3H_6NO^+$		$\rightarrow$	72	1.75
$C_{13}H_{11}^+$	$\rightarrow$ $C_{13}H_{10}^+$	167	$\rightarrow$	166	4.00
	$\rightarrow$ $C_{13}H_9^+$		$\rightarrow$	165	2.00
	$\rightarrow$ $[C_{12}H_{11}]^+$		$\rightarrow$	155	0.20
	$\rightarrow$ $C_{12}H_8^+$		$\rightarrow$	152	1.20
$C_{13}H_{10}^+$	$\rightarrow$ $C_{13}H_9^+$	166	$\rightarrow$	165	45.00
$C_{13}H_9^+$	$\rightarrow$ $C_{13}H_8^+$	165	$\rightarrow$	164	38.57
	$\rightarrow$ $- -$		$\rightarrow$	153	0.86
	$\rightarrow$ $- -$		$\rightarrow$	150	0.75
$C_{12}H_8^+$	$\rightarrow$ $C_{12}H_7^+$	152	$\rightarrow$	151	3.33
	$\rightarrow$ $- -$		$\rightarrow$	141	0.39
$C_3H_6NO^+$	$\rightarrow$ $- -$	72	$\rightarrow$	67	0.90

[a]Bracketed empirical composition data are not supported by high
resolution mass measurements.

[b]Based on a value of 100 for the precursor ion.

NUCLEAR MAGNETIC RESONANCE

NMR STUDIES OF PESTICIDES: AN INTRODUCTION

R. Haque
Department of Agricultural Chemistry and
Environmental Health Sciences Center
Oregon State University
Corvallis, Oregon 97331

The use of nuclear magnetic resonance spectroscopy in solving
pesticide problems has increased considerably in recent years.
There are now two review articles (1,2) available on this subject.
Although the sensitivity of this technique is less than that of
gas-chromatography and mass-spectrometry it offers certain advantages, such as its versatile nature and the ease of interpretation
of the spectra.   NMR has now successfully been employed in studying
the photochemical reaction of pesticides, determining the structure
of pesticides, and monitoring the interaction of pesticides with
biological polymers and surfaces.

The NMR section of this book begins with papers on structural
elucidation of pesticides, their metabolites and related compounds.
The use of shift reagent in the spectral analysis of pesticides is
also illustrated.   NMR characteristics of various polychlorinated
biphenyls have been utilized to study the effect of substituents on
the chemical shift.   The ^{13}C NMR spectra of polychlorinated biphenyls
and other organic compounds are discussed.   The papers of [13]C
reveal its potential in structural elucidation of pesticides.   A
paper describing the basic principles of Fourier Transform (FT)
NMR is also included.   The application of FT NMR to pesticides,
although limited, may be an excellent tool for quantitative analysis
of pesticides.   Since FT method requires a considerably smaller
quantity, the analysis of pesticides at ppm is possible.

The next four chapters deal with the interaction of pesticides
and related compounds with model membranes, proteins and surfaces.
NMR study of the interaction of pesticides with model membranes

has provided a new approach to investigate the molecular basis of mode of action of pesticides. Two introductory papers on the NMR studies involving surface interactions are also presented. Pesticides and related chemicals are adsorbed readily on surfaces and the persistent ones spend a great deal of time in soils and clays. NMR may provide insight on the surface interaction and molecular motion of pesticides, a knowledge currently not available. The last two papers are not on NMR but on closely related topics. One deals with the use of spin-labelling technique to study the interaction of toxicants in vivo whereas the other describes a method of quantitative analysis of pesticides using infra-red spectroscopy.

To date, NMR has served as a powerful tool in physics, chemistry and biology and hopefully it will be found equally useful in studying problems of environmental sciences. Further use of NMR technique, FT method, ^{13}C and other nuclear resonance and relaxation processes may provide valuable basic information on the behavior of toxicants in the environment.

REFERENCES

1. L. H. Keith and A. L. Alford, J. Assoc. Offic. Analyt. Chem., 53, 1018 (1970).

2. R. Haque and D. R. Buhler, 'Nuclear Magnetic Resonance Spectroscopy in Pesticide Chemistry' in "Annual Reports on NMR Spectroscopy" Editor E. F. Mooney, page 232 Academic Press, N.Y. (1972).

NMR CHARACTERIZATION OF SOME CHLORINATED HYDROCARBON PESTICIDES,

THEIR METABOLITES, DERIVATIVES AND COMPLEXES

J. D. McKinney and N. K. Wilson

National Institute of Environmental Health Sciences

P. O. Box 12233, Research Triangle Park, N. C. 27709

and

L. H. Keith and A. L. Alford

Southeast Water Laboratory, WQO, EPA, Athens, Ga. 30601

Abstract. Pmr spectroscopy has been used to assess the stereochemistry of various chlorinated hydrocarbon pesticide metabolites
of the DDT and polycyclodiene type employing essentially all available techniques, such as spin decoupling, selective lanthanide
shift reagents, as well as the Fourier transform system. Both
achiral and chiral shift reagents were particularly helpful in
determining stereoselective and stereospecific chemical and biological transformations of the chlorinated polycyclodiene compounds.
Chemical shift measurements of the benzhydryl and aromatic proton
resonances of various compounds related to DDT (isomers, metabolites
and their derivatives) provided diagnostic stereoelectronic information. Previous work has shown that DDT can undergo two types of
complexation, one involving the benzhydryl proton and trichloromethyl grouping with polar complexing agents, and the other interaction of the DDT aromatic π-electron system with the π-electrons
of a complexing molecule. In respect to these studies an attempt
has been made to classify according to binding type some of the DDT
compounds studied on the basis of the experimental patterns in the
pmr data. Sufficient information is not available at this time to
allow a meaningful correlation to be drawn between binding
propensities and toxicity data.

The role of nmr spectroscopy in the area of pesticide
chemistry has been relatively small as compared to its role in other
areas of chemistry; however, the introduction of Fourier transform
(FT) nmr systems has aroused a great deal of current interest in
this subject by pesticide chemists. The FT system can clearly be
an advantage in the identification of small amounts of pesticide
metabolites and degradation products where comparison standards may
not be available. One can foresee a potential application in pesticide residue analysis to provide more specific and definitive
evidence which may be needed for legal enforcement. In addition,
the FT nmr system is facilitating biological studies of a diverse
nature with nuclei other than hydrogen.[1-5] Nevertheless, pesticide
chemistry has developed to a stage that nmr spectroscopy is not only
used as an instrument for an after-the-fact measurement, but also
can be used as an instrument of prediction. In this regard nmr
spectroscopy has been utilized in our laboratory for structural
elucidation of pesticide conversion products from model complementary chemical reaction studies as well as from metabolic studies.

Specifically, our attention has focused on nmr assessment of
stereochemistry of various chlorinated hydrocarbon pesticide
systems of the DDT and polycyclodiene type employing all available
techniques, such as spin decoupling, selective shift reagents, as
well as the FT system when sufficient amounts of sample are not
available for obtaining spectra in the normal mode. Stereochemical
elaboration was desired since it would facilitate the utility of
these systems as both chemical and biological models for investigating the mechanisms of several enzyme reactions which occur in
the metabolism of these compounds and possibly other related
xenobiotics. Such stereochemical information also relates to
studies of their binding propensities with biopolymers. The polycyclodiene pesticides such as aldrin, dieldrin, and endrin were
particularly attractive models since their metabolisms have been
extensively elaborated,[6] and they constituted quite stable and
rigid models which would facilitate their utility as models and,
at the same time, illuminate the stereospecificities of their
enzymatic reactions. In addition, stereoselective as well as
stereospecific biotransformations can occur in these systems in
which the enantiotopic as well as the enantiomeric properties of
the molecules are manifested.[7]

Experimental Section

All the chemicals used were obtained commercially or synthesized according to their published procedures[8-9] or slight modifications of methods used to prepare similar compounds. The nmr
studies were made with Varian XL-100 Fourier transform (FT), HA-100,
and T-60 spectrometers. Chloroform-d was generally used as the
solvent, but carbon tetrachloride and carbon disulfide were also

used in the complexation studies with DDT. The methods for use of
achiral, Eu(DPM)$_3$, and chiral, Eu(HFC)$_3$, nmr shift reagents with
these compounds to separate superimposed resonances and permit
measurement of optical purity were as previously described.[9] The
temperature measurements for the DDT complexation studies were
made as previously described.[10-11] Spectra were recorded using
the frequency-sweep mode at various sweep times and sweep widths.
Homonuclear decoupling where necessary was done in this mode.
Chemical shifts are expressed in terms of τ values and are relative
to tetramethylsilane (TMS) used as an internal standard.

Results and Discussion

Although some nmr elaboration of the chlorinated polycyclodienes has already been reported in the literature,[12-15] it has
primarily been concerned with the parent pesticides, and in some
cases, there was disagreement among the authors on chemical shift
assignments. In our early studies,[16] our attention was directed
toward clarifying the contradictory aspects of the previous reports as well as providing nmr data on various systems postulated
as metabolites of the polycyclodienes. For the most part, the
metabolite possibilities were obtained by stereoselective syntheses
as pictured in Scheme 1: trans-4,5-dihydroxy-4,5-dihydroaldrin (1),
cis-exo-4,5-dihydroxy-4,5-dihydroaldrin (2), exo-4-hydroxy-4,5-
dihydroaldrin (3), 4-oxo-4,5-dihydroaldrin (4), endo-4-hydroxy-4,5-
dihydroaldrin (5), and cis-exo-4,5-dihydroxy-4,5-dihydroisodrin (6).
Deltaketo endrin (7) was commercially available and was examined
for any similarities its spectra had with those of aldrin ketone (4).
All of these compounds, with the exception of 6, have been impli-
cated as metabolites or degradation products in diverse metabolism
studies.

The detailed synthetic methodology and nmr spectral interpretation of these systems are previously reported.[8] In order to
demonstrate the utility of nmr for assigning stereochemistry as well
as other structural characteristics of these systems, whether they
be of theoretical or metabolic interest, the trans (1) and cis (2)
aldrindiols have been chosen as representative examples. An examination of the pmr spectrum of the trans-diol with the aid of
decoupling (see Figure 1A) reveals several salient features which
reflect the trans-orientation of the hydroxyl groups, viz., the
non-equivalency of all protons except the methylene bridge protons,
the presence of coupling constants involving the C4-C5 protons
which conform well to measured dihedral angles, and the presence
of "W effect" coupling[17] between the C-5 endo-proton and the C-12
anti-proton and the absence of such a coupling with the C-4 proton.
The trans-1,2-relationship is further supported by the addition of
the selective nmr shift reagent,[8,18] tris (dipivalomethanato)-
europium [Eu(DPM)$_3$]. Through complexation equilibria involving the

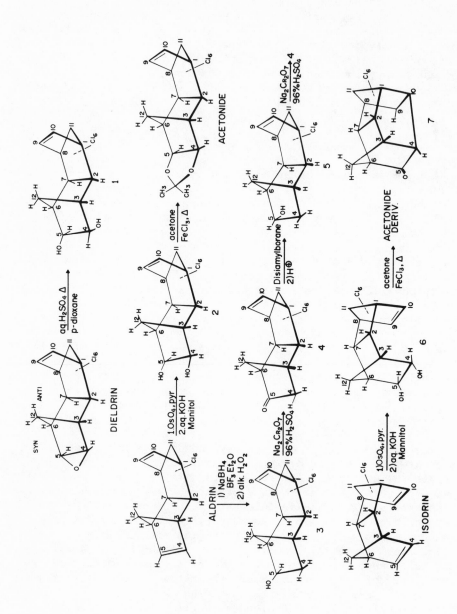

Scheme 1

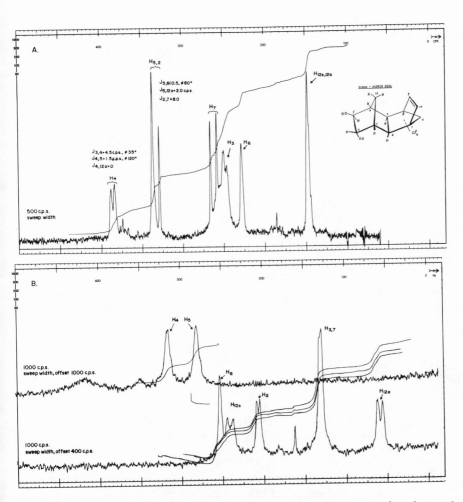

Fig. 1.   (A) <u>trans</u>–aldrindiol (<0.1M) without Eu(DPM)$_3$ and
(B) with 0.2M Eu(DPM)$_3$ in CDCl$_3$ at 100 MHz.

hydroxyl groups, paramagnetic shifts in proton are induced. The
magnitude of the shift is primarily dependent on the distance of
each proton from the center of the europium atom. The addition of
the shift reagent clearly indicates the presence of both exo- and
endo-hydroxyl groups since one observes (see Figure 1B) comparatively
larger deshielding of the C-2 proton by the C-4 endo-hydroxyl than
that observed for the C-7 proton and, likewise, the H_{12s} experiences
larger deshielding than H_{12a} by the exo-hydroxyl association.

The cis-diol (2) pmr spectrum (see Figure 2) also possesses
several characteristics expected of the cis-exo-arrangement of the
hydroxyl groups at C-4 and C-5, viz., the occurrence of equivalent
pairs of protons as a result of the plane of symmetry, a chemical
shift value for the C2-C7 protons expected of exo-hydroxyl groups,
the presence of "W effect" coupling between $H_{4,5}$ and the H_{12a}
proton indicating an endo proton at C4 and C5, and the absence of
any coupling between H_4 and H_5 confirming their equivalency. The
cis-diol was not subjected to nmr analysis in the presence of
Eu(DPM)$_3$ due to its insolubility in suitable solvents. Later nmr
studies utilizing the FT system enabled analysis of the cis-diol
in chloroform-d in the presence of shift reagents. Such studies
as these permit rapid and accurate determination of the relative
stereochemistry without excessive dependence on chemical methods or
spin decoupling experiments.

Recently this nmr approach to assigning stereochemistry of
these systems has been extended for ascertaining stereospecificities
of their enzyme reactions in addition to their stereoselectivities.
Our most interesting result to date has been the nmr confirmation[19]
that the trans-diol can be a product of a stereospecific biological
epimerization of the cis-diol, recently shown[20] in our laboratory
to be a metabolite (in vitro with rat liver tissue) of the polycyclodiene pesticide dieldrin. This was of interest since there is
at least one report[21] that the trans-diol isolated as a metabolite
of dieldrin has optical activity suggesting some stereospecific
process in its formation. Approximately 4 mg of metabolite (trans-
diol) was isolated from the epimerization studies and was then
subjected to nmr analysis, in the presence of a chiral shift
reagent [tris(3-heptafluoropropylhydroxymethylene)-d-camphorato)-
europium (III)], Eu(HFC)$_3$, to measure its optical purity and thereby
possibly confirm the stereospecificity of this biological epimerization reaction. In addition to the biological trans-diol metabolite,
both optically inactive racemic trans-diol and meso-cis-aldrin diol,
prepared as previously reported,[8] were subjected to nmr analysis in
the presence of the chiral shift reagent utilizing the FT nmr
system. The quantity available and the solubility of standard
synthetic trans-diol were sufficient to allow spectra to be obtained
in a single scan in the CW mode. The limited solubility of cis-diol
in chloroform-d and the limited amount of isolable metabolite, however, made it necessary to average 40 to 200 transients in the FT

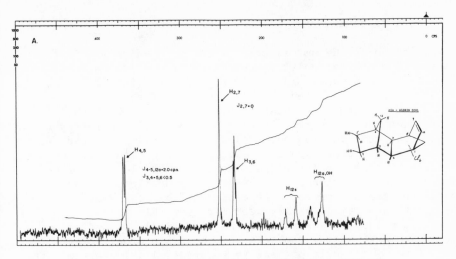

Fig. 2.   Cis-aldrindiol in CDCl₃ at 100 MHz.

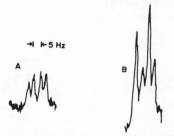

Fig. 3.   (A) C-2 proton signals of racemic <u>trans</u>-aldrindiol-II
(0.23M) with 0.03M Eu(HFC)₃ in CDCl₃ at 100 MHz, Sr/II=0.13; (B)
C-2 proton signals of biological trans-aldrindiol-II (0.035M)
with 0.013M Eu(HFC)₃ in CDCl₃ at 100 MHz, SR/II=0.38.

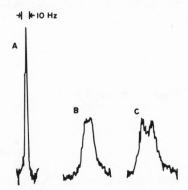

Fig. 4.   C-3C6 proton signals of standard meso-cis-aldrindiol-III
(sat. sol.) with (A) no Eu(HFC)₃; (B) with 0.008M Eu(HFC)₃; (C)
with 0.013M Eu(HFC)₃ in CDCl₃ at 100 MHz.

mode for each spectrum.

Examination of the nmr spectra of the trans-diol (1) in the presence of the achiral shift reagent (see Figure 1B) suggested that the C-2 proton, which is in close proximity (2.2A°) to the endo-hydroxyl group (a potential binding site of the shift reagent), would give the most discernible information with the chiral reagent since its resonance is a doublet with a fairly large coupling constant ($J_{2,7}$=8.0Hz) which is well separated from most interfering proton signals. Spectra of the trans-diol at a molar ratio of chiral shift reagent to diol, (SR/diol), between 0.1 and 0.2 clearly evidence two overlapping doublets, as shown in Figure 3A for the C-2 proton. As expected, the peak areas correspond to an optical purity near zero for this racemic material. Similar nmr analysis of the biological trans-diol likewise gave overlapping doublets for the C-2 proton; however, one enantiomer is clearly present in substantially larger amounts (Figure 3b). The peak areas correspond to an optical purity of 27-29%. Therefore, stereospecificity has indeed been introduced at some stage during the epimerization process. In the absence of shift reagent both standard and biological trans-diol gave essentially identical spectra.

It was of interest to examine the nmr spectra of the symmetrical cis-diol in the presence of the chiral shift reagent since the molecule possesses a reflective plane of symmetry with no rotational symmetry operations. Therefore, each half of the molecule bears an enantiotopic relationship to each other and contains prochiral centers. The C3-C6 protons of the cis-diol were chosen to investigate the effects of the chiral shift reagent since their resonances appear as a broad singlet which is actually a complex multiplet resulting from relatively small coupling. These protons are in fairly close proximity (2.6A°) to the exo-hydroxyl groups. Addition of Eu(HFC)$_3$ to a saturated chloroform-$\underline{d}$ solution of the cis-diol in small increments transforms (Figure 4) the C3-C6 signal from the broad singlet into a pattern which is clearly doublet in character although each component remains broad. This study confirms that each half of the molecule can interact differently with a chiral shift reagent and that such a system could interact with a dissymmetric environment such as an enzyme surface to produce an optically active product.

There is biological evidence[7] that the epimerization process probably involves an α-hydroxy ketone intermediate which can assume exo- and endo-isomeric and enantiomeric forms as well as an ene-diol enantiotopic form. Therefore, it is possible that only one enantiomeric form of the hydroxy ketone is formed in the oxidation step of the epimerization process. On the other hand, only one enantiomeric form of the racemic hydroxy ketone may undergo reduction to the trans-diol. Nmr characterization including the stereochemistry of synthetic aldrin hydroxy ketone is in progress. When these studies are complete and adequate quantities of the hydroxy ketone are available to enable isolation of sufficient amounts of trans-diol from its in vitro biological reduction, optical purity at this stage can be determined.

It is interesting to note that the dieldrin molecule itself has similar properties to the cis-diol in that it possesses a reflective plane of symmetry with no rotational symmetry operations. Therefore, it too contains prochiral centers and would be expected to yield optically active products in biological reactions. Since separate complexing sites are not available in each half of the molecule as in the cis-diol, enantiotopic properties cannot be demonstrated by the nmr technique. Similarly, the DDT molecule has these symmetry properties assuming that restricted rotation does not preclude a mirror image plane in the preferred rotamer. On the other hand, restricted rotation in the DDT molecule may enable it to exist in enantiomeric forms. NMR evidence for restricted rotation in DDT will be discussed later.

Nmr analysis of another dieldrin metabolite, viz., the major fecal metabolite in the rat, has been done utilizing the achiral shift reagent (Eu(DPM)$_3$) again to separate superimposed resonances. This is an example of where no synthetical comparison standard was available and nmr interpretation rests, to a large extent, on previous knowledge of the nmr spectra of analogous compounds. There was some disagreement in the literature[22] on the structural assignment for the metabolite; however, batch metabolic production of the material enabled more interpretable and definitive data to be obtained. Some of the confusion was centered around an apparent AB quartet found (Figure 5) in the metabolite nmr spectrum between τ5.9 and 6.3 similar in coupling constant to the AB quartet observed for the C-12 geminal protons of dieldrin but markedly deshielded (C-12 protons of dieldrin between τ8-9). This was later shown to be due to a rather strong and unusual vicinal coupling between the C-12 hydroxyl proton rigidly held by the hydrogen bonding to the epoxide oxygen and the C-12 anti-proton. This was established by D$_2$O acid catalyzed exchange experiments as well as the addition of Eu(DPM)$_3$.

Without elaborating the data again (previously reported[22]), pmr analysis of this metabolite provided the following information. Decoupling experiments indicated coupling constants of smaller magnitude for the epoxide protons and the adjacent bridge head protons as compared to the same couplings in dieldrin. This reflects steric compression of the C-12 hydroxyl group against the electronic clouds of the epoxide group. In addition, irradiation experiments indicated the presence of a "W effect" coupling of the epoxide protons and the C-12 anti proton, thereby confirming the syn relationship of the hydroxyl group to the epoxide group. The

addition of Eu(DPM)$_3$ destroyed the AB quartet and, therefore, confirmed hydrogen bonding since the rigidity of the hydrogen bond
would be effectively weakened by the competitive Lewis acid
properties of the shift reagent for the lone pair of electrons on
the epoxide oxygen. In addition, the shift reagent study clearly
revealed the presence of three equivalent pairs of protons as
expected for the symmetrical structure. Even though a comparison
standard was not available, this nmr data along with other spectral
and chemical data allows unequivocal assignment of the structure
of this major dieldrin metabolite.

In connection with recent metabolic studies of dieldrin,
chemical investigation of the requirements for involvement of the
C-12 methylene and dichloroethylene bridges of the molecule has
been undertaken. This was of interest since the bridging process
is unique to the metabolism of dieldrin of the endo-exo skeleton,
and the products formed seem to be more toxic than their parents
and bear some stereochemical resemblance to cyclosteroids. There
has been at least one report[23] of a chemically induced involvement
of these bridges yielding a product partially characterized by nmr.
The bridging process can occur readily under photolytic conditions
yielding a variety of products, depending on starting material and
conditions, some of which have been partially characterized by nmr
analysis.[12,24-26] NMR elaboration of photodieldrin and related
bridged systems obtained from varied chemical and photolytically
induced reactions is in progress including the measurement of
coupling constants with the aid of shift reagents.

In addition to the application of nmr shift reagents in
structural and conformational studies, other applications involve
competition experiments to possibly measure isomer ratios, etc.,
and still others involve the use of relative shifts to obtain distance information. Since lanthanide ions can frequently substitute
for Ca^{+2} and Mg^{+2} while retaining biological activity, the possibility exists for investigating the shapes and sizes of active
sites in biological systems.

The high resolution nuclear magnetic resonance spectra of the
DDT class of pesticides and related compounds have been previously
reported,[27] including a study of the resonances of the aromatic
protons as they are affected by various substituents. The present
discussion of their spectra will focus on the stereochemical information which can be obtained from their nmr analysis which may
have some biological significance in possibly effecting differences
in binding and/or metabolism propensities. Of particular interest
to us was the effects of various functional groups on the benzhydryl proton of bis(p-chlorophenyl) systems since this was a
measure of the chemical environment of a potential binding site in
the DDT molecule. Table I lists the benzhydryl proton shifts for
several model systems, some of which occur as metabolites of DDT.

Table I.   Chemical shifts (τ) of benzhydryl proton of (p-chloro-phenyl)$_2$ CHR for various functional groups (R).

R	Compound	τ
CCl$_3$	1,1,1-trichloro-2,2-bis(p-chlorophenyl)-ethane, DDT	5.00
CF$_3$	1,1,1-trifluoro-2,2-bis(p-chlorophenyl)-ethane, F-DDT	5.34
CHCl$_2$	1,1-dichloro-2,2-bis(p-chlorophenyl)ethane, DDD	5.49
CH$_2$Cl	1-chloro-2,2-bis(p-chlorophenyl)ethane, DDMS	5.75
CH$_2$OH	2,2-bis(p-chlorophenyl)ethanol, DDOH	6.00
CHO	2,2-bis(p-chlorophenyl)acetaldehyde, DDCHO	5.08
COOH	2,2-bis(p-chlorophenyl)acetic acid, DDA	5.03
COOCH$_3$	methyl bis(p-chlorophenyl)acetate, DDA methyl ester	5.06

A study of the chemical shift values for these compounds reveals some interesting stereoelectronic information. The tri-chloromethyl grouping clearly exerts a rather strong deshielding effect upon the benzhydryl proton. Similar effects are apparent in the three carbonyl containing functional group compounds which is probably a result of the paramagnetic anisotropy of the carbonyl moiety. Therefore, a similar chemical environment to DDT at this position is suggested without the steric requirements of the tri-chloromethyl grouping. A possible measure of the deshielding effects due to the steric requirements and proximal nature of the chlorine atoms can be seen in a comparison with the trifluoro analog of DDT. Fluorine is the most electronegative of halogens but has an atomic radius nearly equal to that of hydrogen. In addition, it has no d-orbitals and therefore cannot enter into resonance with π-electron systems. It can be seen that even though an upfield shift might result from the increased electronegativity of the fluorine atoms and hyperconjugation, the difference in the two values ($\Delta\delta 0.34$) seems too large to account for this effect alone. As chlorine atoms are removed the deshielding is reduced until it is essentially eliminated in the saturated alcohol compound (DDOH). Space filled models show that with the loss of only one chlorine atom of the trichloromethyl group as in DDE, a dehydro-chlorination product of DDT, the strain and restricted rotation of the DDT molecule can be relieved. The overwhelming effects of the trichloromethyl group are again apparent, if one compares the benzhydryl proton resonance value (τ5.06) for methoxychlor [1,1,1-trichloro-2,2-bis(p-methoxyphenyl)ethane] with the same resonance found in DDT. The benzhydryl proton of methoxychlor is much less acidic than the same proton in DDT since DDT undergoes base catalyzed dehydrochlorination at a rate near 250 times that of methoxychlor. This would imply an increased electron density (shielding) about the benzhydryl carbon in methoxychlor due to the electron donating resonance effects of the p-methoxyl groups; however, this is not evident from a measure of the benzhydryl proton chemical shift ($\Delta\delta$0.06) suggesting that the trichloromethyl group exerts an overriding electronic effect on this position which for reasons that are unclear, is not translated into a similar rate for the base catalyzed dehydrochlorination process.

The deshielding effect of the CCl_3 group is most pronounced when chlorine substitution on one of the aromatic rings occurs in the ortho position. The enhanced deshielding effect is evident from a comparison of the benzhydryl proton resonances of p,p'-DDT versus o,p'-DDT, τ5.00 as compared to τ4.25. Taking advantage of this difference in chemical shift, one can separately integrate the two resonances and, therefore, determine the percentage of each isomer present in a mixture. Figure 6 shows the nmr spectrum of technical DDT which consists of 80% p,p'-DDT and 20% o,p'-DDT.

Similar effects of substituents on C_α and C_β are observed on the ortho and, to some extent, the meta aromatic protons flanking the para substituted chlorines. Table II lists the τ values of the ortho and meta protons for a representative series of p-chloro substituted systems.

Table II. τ Values of various ortho and meta substituted protons flanking a para chlorine, $(p-Cl\emptyset)_2 - C \begin{smallmatrix} R_1 \\ R_2 \end{smallmatrix}$

Compound	R_1	R_2	H_o	H_m
DDT	H	CCl_3	2.56	2.74
DDT (trifluoro)	H	CF_3	2.65	2.65
DDD	H	$CHCl_2$	2.81	2.75
DDE	–	$=CCl_2$	2.83	2.73
DDOH	H	CH_2OH	2.97	2.81
DDA	H	COOH	2.77	2.71
kelthane	OH	CCl_3	2.40	2.76

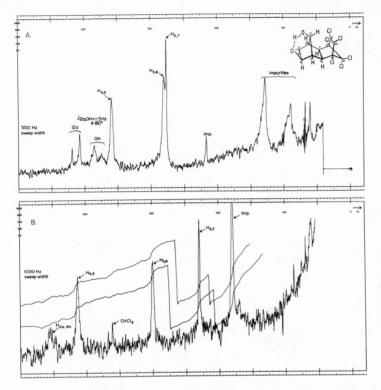

Fig. 5. (A) Dieldrin Fecal Metabolite (~5mg/0.1ml) without Eu(DPM)$_3$ and (B) with Eu(DPM)$_3$ in CDCl$_3$ at 100 MHz.

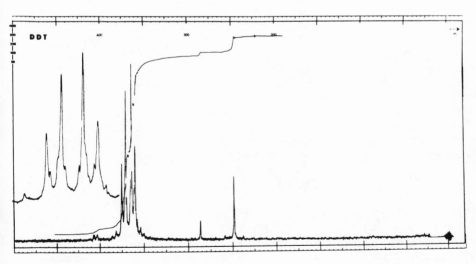

Fig. 6. PMR spectrum of technical DDT, a mixture of o,p'-DDT and p,p'-DDT, in CDCl$_3$ at 100 MHz.

A study of the values for the <u>ortho</u> proton resonances indicates that the CCl$_3$ moiety again exerts the greatest deshielding influence which is enhanced when in combination with a hydroxyl group. The CF$_3$ moiety appears to have little or equal influence on the aromatic protons since both the <u>ortho</u> and <u>meta</u> protons appear to be equivalent. The apparent deshielding (over that in, e.g., DDD) for the trifluoro DDT aromatic protons is not steric in nature but probably due to the subtle electronic effects of the CF$_3$ moiety such as a field effect. The CF$_3$ moiety can be a unique functional group in a molecule. As chlorines are removed and replaced by hydrogen, the deshielding effects are reduced as was noted for their benzhydryl protons. It is interesting that in DDE and DDD the <u>ortho</u> and <u>meta</u> protons are experiencing essentially identical chemical environments suggesting no appreciable π-electron involvement of the aromatic ring and the DDE double bond. Space filled models support this argument since the system cannot easily attain planarity of aromatic rings with the double bond to allow overlap of orbitals. As expected, the saturated alcohol shows the least deshielding. The aromatic protons of the carboxylic acid are nearing equivalency with moderate deshielding (relative to DDOH) as a result of the electronic properties of the carbonyl moiety. In general, the <u>meta</u> protons are not that different except in DDOH where no chlorine is present in the ethane portion of the molecule. When groups other than chlorine are substituted in the <u>para</u> position of the aromatic ring, shifts in the <u>meta</u> proton resonances can be observed. For example, a <u>para</u> ethyl substituent (Perthane) shifts it to τ2.92 and a methoxy (methoxychlor) group shifts it even further upfield to τ3.18.

It is interesting to compare the <u>ortho</u> proton resonance in the aromatic ring of o,p'-DDT containing the <u>ortho</u> substituted chlorine with the <u>ortho</u> proton of the p-substituted ring. This resonance falls at τ1.88 as compared to 2.52 (in p,p'-DDT, τ2.56) for the <u>ortho</u> protons in the ring containing the <u>para</u> chlorine. Figure 7 shows the nmr spectra of p,p'-DDT and o,p'-DDT with an expansion of their aromatic regions. Space filled models indicate restricted rotation about the aromatic-aliphatic C-C bond in the o,p'-DDT molecule. The most favorable rotamer places the <u>ortho</u> proton close to the β-carbon chlorines and the <u>ortho</u> chlorine as far as possible from the CCl$_3$ moiety. Such a situation would account for the unusual deshielding observed for the <u>ortho</u>-proton in o,p'-DDT.

NMR has also been utilized for the structural elucidation of various chemical derivatives containing the <u>bis</u>(p-chlorophenyl) grouping, some of which were theoretical models for investigating DDT metabolism. Of particular interest was the chemistry encompassing the possible intermediacy of 2,2-<u>bis</u>(p-chlorophenyl)acetaldehyde (DDCHO) in DDT metabolism. The results of these chemical studies have been previously reported.[28] NMR has confirmed the facile enolization of this aldehyde under certain conditions. An enol ether acetate derivative studied, $(P-Cl\phi_2)_2C = CHOCH_2CH_2OAc$, was an interesting model since it undergoes thermal and acid catalyzed rearrangement ultimately leading to 4,4'-dichlorobenzophenone (DBP). DBP is a known DDT metabolite and has been shown to be a common decomposition product of the aldehyde, DDCHO, and its derivatives. An oxonium resonance contribution characteristic of enol ethers would be expected to place some negative charge on the benzhydryl carbon; however, nmr analysis at ambient temperatures did not support this since its aromatic resonances (multiplet, centered at $\tau 2.76$) were not appreciably different from those of, e.g., DDE ($\tau 2.73$–2.83). On the contrary and as predicted from the mechanism of rearrangement, the carbon must carry some partially positive charge since it readily undergoes intramolecular nucleophilic attack at this position to produce oxygenated benzhydryl derivatives. These derivatives can readily convert to the benzophenone, e.g., kelthane is known to decompose to the benzophenone and chloroform. Therefore, in a model situation which should ideally place negative charge at the benzhydryl carbon through resonance, one observes the opposite tendency.

A knowledge of the stereoelectronic properties inherent to DDT and related molecules is essential to gaining an understanding of the exact types of molecular complexation that these systems experience in biological systems. Such complexing interactions seem important to delineating the mechanism and mode of action of DDT and related pesticides. The nmr data described has increased our knowledge of the peculiar stereoelectronic properties of these systems in solution; however, this stereochemistry does not necessarily have to be that required in a biological interaction.

NMR spectroscopy has been utilized[10-11,29-30] to directly study the behavior of model complexes of DDT and a few related systems in solution since it should enable more precise characterization of the possible molecular association phenomena involved. It was felt that once the types of binding to biologically significant compounds were established, structurally different (stereoelectronically and otherwise) analogs of DDT (some of which may be DDT metabolites) could be used to test the proposed binding mechanisms. Ultimately, one would like to show a correlation between the differences in binding propensities and the toxicity of the DDT-type compound.

NMR studies of the complex equilibria of DDT have shown[10] that two types of complexes are formed, one association involving primarily the benzhydryl trichloromethyl grouping with a polar complexing agent (Type I), and the other interaction of the DDT aromatic π-electron system with π-electrons in the complexing molecule (Type II). Type I complexation was measured directly from the nmr spectrum as a downfield shift (deshielding) of the benzhydryl proton, and Type II complexation as an upfield shift (shielding) of the meta aromatic protons (ortho to para chlorine substituent), which are

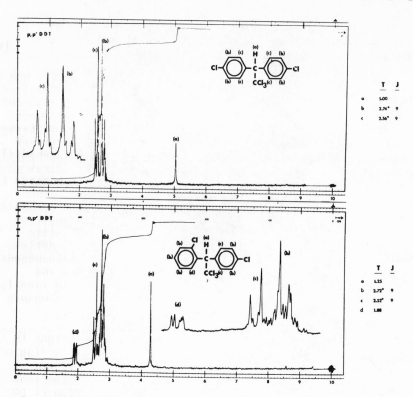

Fig. 7.   PMR spectra of p,p'-DDT and o,p'-DDT with an expansion
of their aromatic regions.

somewhat more difficult to measure, or as an upfield shift in the
benzhydryl proton. All three sets of protons in DDT show a linear
dependence of their chemical shifts on the concentration of complexing agent. Type I complexation was the weakest of the two and was
strongest for polar complexing agents in which polarization of the
molecule could be supported by resonance. Representative examples
of this type of polar complexing reagent are ethyl ethylcarbamate,
tri-n-butyl phosphate, ethyl acetate, and N,N'-dimethyl urea in
order of decreasing binding strength. Figure 8 shows the linear
relationship obtained when data are plotted for the observed benzhydryl proton shift divided by the concentration of complexing
agent as a function of the observed benzhydryl proton shift for
DDT-ethyl ethylcarbamate complexation at 40.0°. Other types of
binding involving the benzhydryl proton such as hydrogen bonding
do not appear to be an important means for DDT complexation.

Type II complexation was somewhat stronger (2 to 3 times) than
Type I and appears to be significant only for π-electrons of
aromatic rings and not π-electrons of isolated or otherwise con-
jugated double bonds. Examples of this type of aromatic complexing

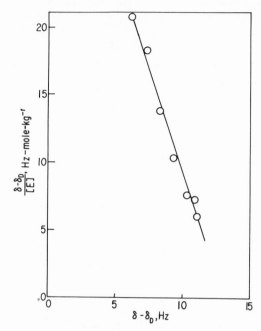

Fig. 8. The observed benzhydryl proton shift divided by the concentration of complexing agent as a function of the observed benzhydryl proton shift for DDT-ethyl ethylcarbamate complexation at 40.0°C.

agent are naphthalene, benzene, and indole in order of decreasing binding strength. The aromatic complexing agents are the only ones which gave any evidence of forming complexes of 1:2 as well as 1:1 stoichiometry. Figure 9 shows a plot of the observed benzhydryl proton shift as a function of the concentration of complexing agent for DDT-indole complexation at 16.0° which takes into consideration both 1:1 and 1:2 equilibria. Figure 10 illustrates the dependence of the logarithms of the equilibrium constants on the reciprocal of the absolute temperature for DDT-indole complex formation. Handling of the data in this manner allows calculation of equilibrium constants and values for standard thermodynamic parameters. Generalized solvent effects do not account for the much larger concentration dependencies of chemical shifts observed with the complexing agents. Figure 11 illustrates the dependence of chemical shift of the DDT meta aromatic protons on the concentration of several complexing agents. Only a slight dependence of the meta protons on the concentration of added n-hexane was observed.

Type I binding appears to be in some ways more favorable than Type II binding since there are fewer steric constraints in the early stages of molecular association. Both ester groupings and phosphate groupings are potential binding sites in biopolymers for

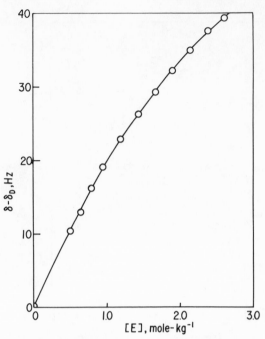

Fig. 9.   The observed benzhydryl proton shift as a function of
the concentration of complexing agent for DDT-indole complexation
at 16.0°C.

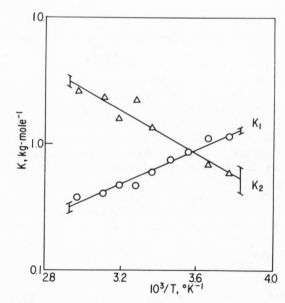

Fig. 10.   The dependence of the logarithms of the equilibrium
constants on the reciprocal of the absolute temperature for DDT-
indole complex formation.   Brackets indicate the standard error.

this type of complexation. For example, DDT has been shown[29] to
bind with the phospholipid lecithin. Type II binding would be more
important for aromatic constituents in protein such as phenyl-
alanine or tryptophan, in preference to binding at amide peptide
linkages. Complexation of this type with dissymmetric enzymes
could be induced by the possible prochiral nature of the aromatic
rings of DDT and certain other related systems. Type I binding
seems to be largely dependent on the availability of an acidic
proton and not so much on other peculiar structural characteristics.
The acidic nature of the benzhydryl proton is due largely to the
negative inductive effect and proximity of the trichloromethyl
group and, in part, to electronic involvement with the π-electrons
of the aromatic rings (an allylic coupling between the benzhydryl
proton and the _ortho_ aromatic protons has been observed).[10]

It would be interesting to measure the binding parameters of
the trifluoromethyl analog of DDT since it would be expected to
undergo strong Type I binding but weak Type II binding. In addition, both DDA and methoxychlor should undergo weaker Type II
binding (relative to DDT) whereas methoxychlor appears to be a
much weaker Type I binder than DDA. On the other hand, kelthane
may show some rather interesting Type II binding properties due to
the extra steric strain and restricted rotation imparted by the
hydroxyl group in addition to its apparently more electron deficient
aromatic rings. Kelthane acetate (_ortho_ aromatic protons, $\tau 2.40$)
would also be an interesting model since, in addition to increased
steric requirements, one would expect an appreciable effect on the
aromatic rings by the tendency of the molecule to lose an acetoxy
group as shown by previous work.[31] o,p'-DDT would be expected to
be a poor Type I binder due to the proximity of the β-chlorine to
the benzhydryl proton in the most stable conformer. Loss of at
least one of the β-chlorines (o,p'-DDD) should relieve this effect
and restore Type I binding potential. Table III summarizes the
major features of the complexation data found for DDT in our laboratory. Item 5 (for each Type) lists the strongest binding possibilities which are suggested by the chemical shift data in Tables
I & II. A few of these DDT systems have been tested[29] for their
ability to undergo Type I binding with the phospholipid lecithin,
and some degree of correlation with toxicity data exists.

Complementary C^{13} nmr studies would provide additional information on the electronic properties of DDT compounds, and hence,
their binding propensities, since such studies would reflect a more
accurate picture of the total charge densities at the carbons
involved. Such additional nmr studies of DDT and its derivatives
in conjunction with the many extant studies of their metabolism and
toxicity should provide a basis for evaluating the validity of
correlating experimental trends in nmr data and binding with the
various toxicological measurements.

Table III.   Complexation types of DDT compounds.

Type I - Benzhydryl-trichloromethyl
         Association measured as downfield shift
         (deshielding) of the benzhydryl proton

1. Weaker of the two types.

2. Strongest for polar complexing agents in which polarization of
    the molecule can be supported by resonance, e.g., ethyl
    ethylcarbamate.

3. Fewer steric constraints in the early stages of molecular
    association.

4. Reflects acidic nature of the benzhydryl proton imparted by
    negative inductive effect and proximity of trichloromethyl
    group, and, in part, to electronic involvement with π-electrons
    of the aromatic rings.

5. Benzhydryl proton shifts indicate that p,p'-DDT, DDA, DDA-CH_3
    ester, DDCHO, p,p'-DDD, and trifluoro-DDT are the strongest
    binding possibilities.

6. Ester and phosphate groupings are potential binding sites in
    biopolymers.

Type II - Aromatic π-electron system with π-electrons in
          complexing molecule measured as upfield shift
          (shielding) of the meta aromatic protons or
          benzhydryl proton

1. Stronger of the two types.

2. Significant for π-electrons of aromatic rings and not π-electrons
    of isolated or otherwise conjugated double bonds, e.g., indole.

3. Can form complexes of 1:2 as well as 1:1 stoichiometry with
    more stringent steric requirements.

4. Reflects electron accepting ability of DDT π-electron system im-
    parted by negative inductive and field effect and proximity of
    trichloromethyl group.

5. Aromatic proton resonances indicate that p,p'-DDT, o,p'-DDT,
    kelthane, and kelthane acetate are the strongest binding
    possibilities.

6. Aromatic constituents in protein and purines and pyrimidines in
    nucleic acids are potential binding sites.

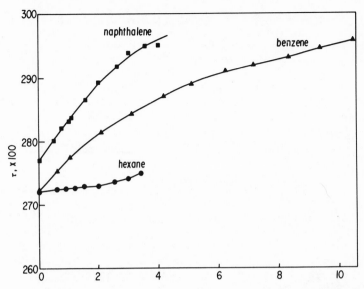

Fig. 11. Dependence of the chemical shift of DDT aromatic protons
3 and 4 on the concentration of several possible complexing agents
at 27.0°C. The curves shown for benzene and naphthalene are the
calculated curves from non-linear regression analysis using Eq. (4).

NMR spectroscopy has been an invaluable tool enabling the
assignment of relative stereochemistry as well as measurement of
optical purity of metabolites and prediction of stereochemical
behavior of a variety of chlorinated hydrocarbon pesticide systems
of the DDT and polycyclodiene type. There is no other single
method for studying these systems which can provide as much information. The availability of the FT system will increase the usage
of nmr in all of the capacities discussed for continuing studies
of these and other environmentally important compounds.

References

1.   O. Jardetzky and N. G. Wade-Jardetzky, Ann. Rev. Biochem., 40,
     605 (1970) for a general review of nmr applications to biolog-
     ical problems.
2.   R. H. Sarma and R. J. Mynott, Org. Magn. Resonance, 4(4), 577
     (1972).
3.   S. Cheng and M. Martinez-Carrion, J. Biol. Chem., 247(20),
     6597 (1972).
4.   A. S. Perlin, N. M. Kin, Y. Ng, S. S. Bhattacharjee and L. F.
     Johnson, Can. J. Chem., 50(15), 2437 (1972).

5. D. G. Davis and G. Inesi, Biochim. Biophys. Acta, **282**(1), 180 (1972).

6. G. T. Brooks, In "Chlorinated Insecticides", CRC Press, Cleveland, Ohio, 1973, pp. 300.

7. J. D. McKinney, S. M. dePaul Palaszek, H. B. Matthews and H. Mehendale, Pesticide Symposia Proceedings, Vol. II, 8th Inter-American Conf. Toxicol. & Occup. Med., July, 1973 (in press).

8. J. D. McKinney, L. H. Keith, A. Alford and C. E. Fletcher, Can. J. Chem., **49**(12), 1993 (1972).

9. J. E. Peterson and W. H. Robison, Toxicol. Appl. Pharmacol., **6**, 321 (1964).

10. N. K. Wilson, J. Am. Chem. Soc., **94**, 2431 (1972).

11. N. K. Wilson and W. E. Wilson, Sci. Total Environ., **1**, 245 (1972).

12. A. M. Parsons and D. J. Moore, J. Chem. Soc., (C), 2026 (1966).

13. A. P. Marchand and J. E. Rose, J. Am. Chem. Soc., **90**(14), 3724 (1968).

14. J. A. Bukowski and A. Cisak, Ann. Soc. Chim. Polonorum, **42**, 1339 (1968).

15. R. McCullock, A. R. Rye and D. Wege, Tetrahedron Letters, **59**, 5163 (1969).

16. L. H. Keith, A. L. Alford, and J. D. McKinney, Tetrahedron Letters, **28**, 2489 (1970).

17. J. Meinwald and A. Lewis, J. Am. Chem. Soc., **83**, 2769 (1961).

18. J. K. M. Sanders and D. H. Williams, Nature, **240**, 385 (1972) for a review of the use of shift reagents in nmr.

19. J. D. McKinney, H. B. Matthews and N. K. Wilson, Tetrahedron Letters (1973) (in press).

20. H. B. Matthews and J. D. McKinney, Presented at the 5th International Congress on Pharmacology, July, San Francisco, Calif., 1972.

21. L. Korte and H. Arent, Life Sciences, **4**, 2017 (1965).

22. J. D. McKinney, H. B. Matthews and L. Fishbein, J. Agric. Food Chem., **20**(3), 597 (1972) and references therein.

23. A. S. Y. Chau and W. P. Cochrane, Bull. Environ. Contam. Toxicol., **5**(6), 515 (1971).

24. L. H. Keith, A. L. Alford and J. D. McKinney, Analytica Chimica Acta, **60**, 1 (1972).

25. L. H. Keith, Tetrahedron Letters, **1**, 3 (1971).

26. D. Bieniek and F. Korte, Tetrahedron Letters, **46**, 4059 (1969).

27. L. H. Keith, A. L. Alford and A. W. Garrison, Journal of the Assoc. of Offic. Anal. Chem., **52**(5), 1074 (1969).

28. J. D. McKinney, R. E. Hawk, E. L. Boozer and J. E. Suggs, Can. J. Chem., **49**(23), 3877 (1971).

29. R. Haque, J. Tinsley and D. Schmedding, Mol. Pharm., **9**(1), 17 (1973).

30. R. T. Ross and F. J. Biros, Biochem. Biophys. Res. Commun., **39**, 723 (1971).

31. J. D. McKinney and L. Fishbein, Chemosphere, **1**(2), 67 (1972).

'H Nmr Spectra and Structure of a Unique Tetrachloropentacyclo-

dodecanone Derived from Monodechloroaldrin and Monodechloroisodrin

K. Mackenzie,[*] C.H.M. Adams and D.J. Cawley

School of Chemistry, University of Bristol, England

Some time ago we found[1] that the hexachlorocyclodiene
pesticides aldrin (1) isodrin (2) dieldrin (3) and endrin (4)
(and several related compounds) were stereo-selectively mono-
dehalogenated at the dichloromethylene bridge by heating with
the elementary reagent Na(K)OR-DMSO (with or without added ROH)
giving e.g. from aldrin and isodrin the pentachloro compounds
(5), (6).   The dienes (5) and (6) behave as expected when
treated with acetic-sulphuric acid mixtures, dechloroaldrin giving
almost entirely unrearranged acetate (7) and ca. 1% of a tetra-
chlorocarbonyl compound ($\sqrt{}$max 1785 vs cm.$^{-1}$ m$/\overline{e}$ 310) whilst
dechloroisodrin, characteristically, gives predominantly the
half-cage acetate (8).[1]

(1) R = Cl
(5) R = H
(3) R = Cl, exo-epoxide

(2) R = Cl
(6) R = H
(4) R = Cl, exo-epoxide

The propensity of isodrin (and endrin) to give half-cage compounds
related to structure (8) when treated with electrophilic reagents
is well known:[2] on the other hand, aldrin is relatively inert
to skeletal rearrangements, although some conversion to compounds
directly derived from the isodrin skeleton has been observed.[3]
The behaviour of carbocations derived from these halogenated
structures thus contrasts with that of the analogous stereoisomeric
hydrocarbon ions derived solvolytically, which undergo multiple
skeletal rearrangements to give a variety of common products.[4]
It was therefore of considerable interest to us when we found that
the pentachlorodienes (5) and (6) are hydrolysed in conc. sulphuric
acid (with or without an inert solvent) to give the same initial
tetrachloroketone (74% yield, m.p. 150–152°, ν_{max} 1785 vs cm.$^{-1}$
m/e 310) resulting formally from protonation, ring closure,
hydration and loss of HCl.   Several minor common reaction products
are also formed (each $\ngtr$ 0.3%).   The ketonic product – shown to
have structure (9) – isomerises when chromatographed on silicagel
(petroleum-dichloromethane) and incorporates a single atom of
deuterium when similarly isomerised in weakly basic MeOD (m/e 311,
τ 5.44 absent).   Thus the isomeric ketone(10) (m.p. 165–167°)
appears to be the epimer of an α-chloroketone, (9), virtually
eliminating the "staggered" half-cage ketone (11) as a possible
structure.   Simpler half-cage ketones such as (12) had been
isolated in our earlier work,[1] and neither this ketone nor related
structures[2,5] have 'H nmr spectra remotely resembling that of
ketones (9) and (10).

(7)   (8)

(9) R^1=Cl R^2=H
(10) R^1=H R^2=Cl

(11)   (12)

The ^{1}H nmr spectra of ketones (9) and (10) are in fact complex and
unusual (Fig.1,2); chemical shift data, coupling constants and
signal complexity when taken together with double resonance
experiments harmonise however with the structures illustrated for
these ketones.   The large chemical shift difference ($\Delta\tau$ = 1.75)
(Table 1) for endo- and exo-12-H can be understood on the basis
of severe steric interaction between the former proton and chlorine
on C-3 (which approach to within less than the sum of the
van der Waals radii): thus the endo proton is strongly deshielded
and the exo proton shielded - as has been found in other cases,
e.g. for proton-proton interaction.[6]   Similar chlorine proximity
deshielding is observed for H-8 in ketone (10) ($\Delta\tau$ = 0.4), and for
H-4 ($\Delta\tau$ = 0.3) relative to the chemical shifts for these protons
in ketone (9).

Further confirmation of the proposed structures (9) and (10)
for the ketone isomers, and of the mechanistic aspects of their
formation by sequential Wagner-Meerwein rearrangements is derived
by carrying out the hydrolyses of dienes (5) and (6) in 98%
D_2SO_4-D_2O (99.5% isotopically pure).   The nmr spectra of the
products of deuterolysis of the two dienes (Fig. 3,4) are clearly
different: the ketone derived from the diene (5) loses the AB
quartet characteristic of the C-10 methylene bridge, this signal
collapsing to a broadened singlet, exactly as would be expected
for initial deuteronation followed by Wagner-Meerwein rearrangement
to the same (or closely similar) ion as formed initially from
diene (6) (see Scheme 1). The deuterio ketone formed from diene (6),
as expected, lacks the highly characteristic upfield quartet due to
exo-12-H, the endo proton doublet of multiplets collapsing to a
broadened singlet (partially obscured).

The ^{1}H nmr spectra of the ketones can be further analysed
by introducing deuterium at C-10 and C-11 making use of keto-

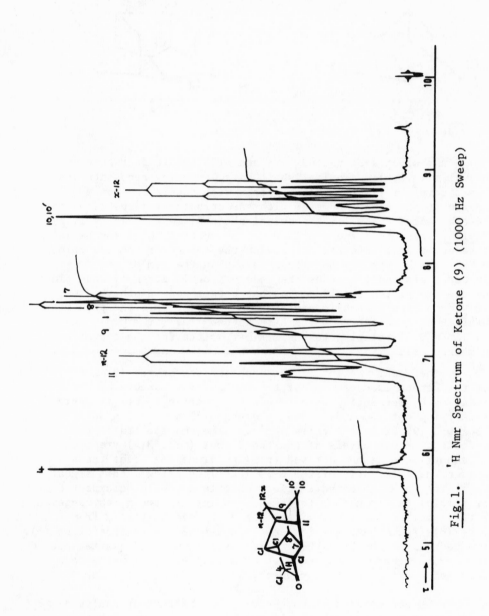

Fig. 1. ^{1}H Nmr Spectrum of Ketone (9) (1000 Hz Sweep)

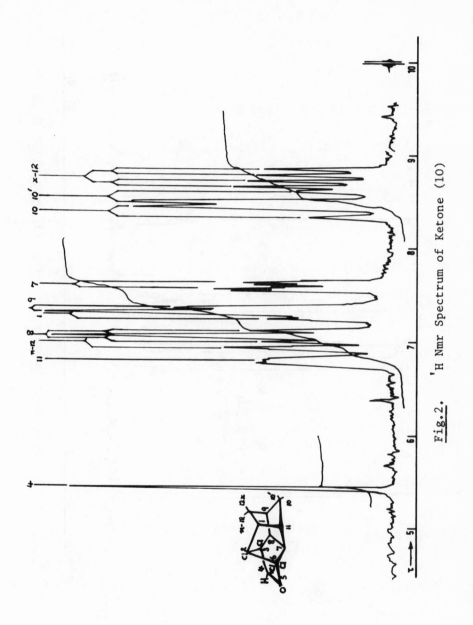

Fig.2. ¹H Nmr Spectrum of Ketone (10)

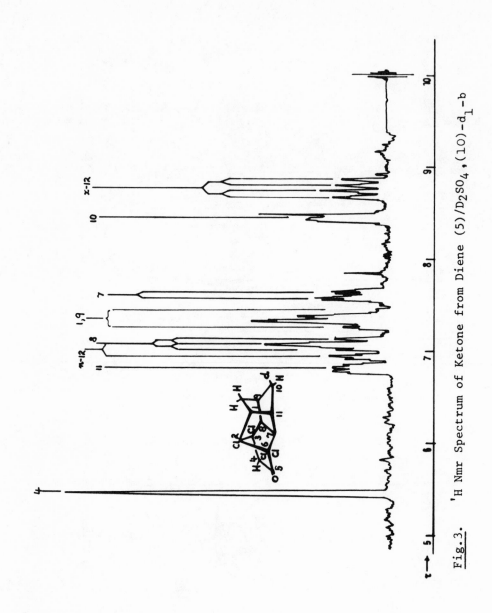

Fig.3. ^{1}H Nmr Spectrum of Ketone from Diene (5)/D_2SO_4,(10)$-d_1$-b

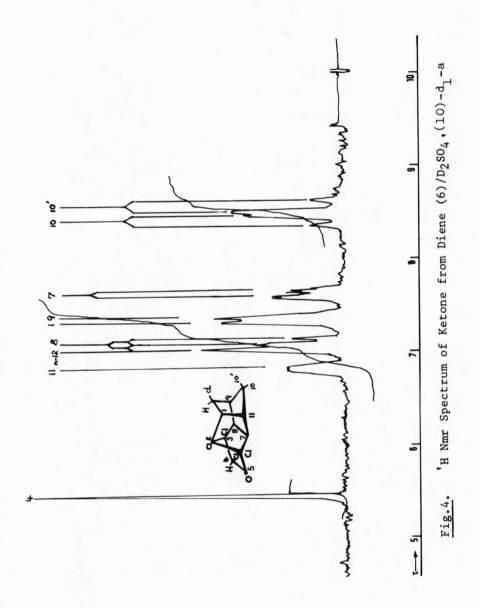

Fig.4.   ¹H Nmr Spectrum of Ketone from Diene (6)/D$_2$SO$_4$,(10)-d$_1$-a

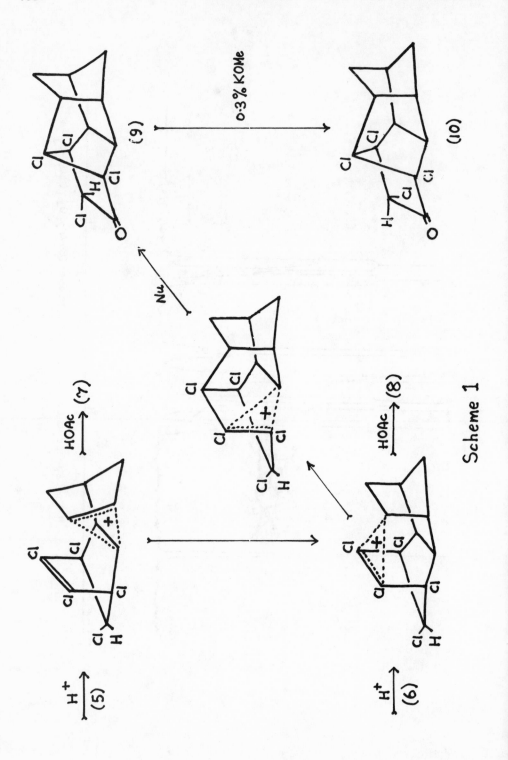

Scheme 1

dechloroaldrin[7] (15), in which the α-methylene protons are readily exchanged for deuterium; subsequent reduction with $LiAlH_4$ or $LiAlD_4$ then affords the bis- or tris- deuterio alcohols (16)-d_2, (16)-d_3.

(9)-d_2 $R^{1,2}$ = D ⟶ (10)-d_2
(9)-d_3 R^{1-3} = D ⟶ (10)-d_3

Preliminary experiments had indicated that treatment of alcohol (16) with sulphuric acid affected entry into the same "nest" of cations as protonation of diene (5) and that the deuterated alcohols were thus suitable precursors of the labelled ketones (9)-d_2, (9)-d_3 and their C-4 epimers (10)-d_2, (10)-d_3. The ^{1}H nmr spectra of these labelled ketones (Fig. 5,6,7) are entirely consistent with the proposed structures, the C-10 AB quartet being eliminated in (9)-d_2 and (10)-d_2 and in addition the broad complex multiplet due to H-11 being absent in (10)-d_3. Other signals in the spectra are simplified and/or sharpened in the expected way.

Epimerisation at the chloromethylene group in the ketonic hydrolysate does not entirely rule out structure (11) for the ketone: however reduction of the ketone (LiAlH$_4$) gives the expected endo/exo mixture of alcohols whose ^{1}H nmr spectrum shows the anticipated pairs of doublets due to cis amd trans ClCH.CHOH, with the minor exo-alcohol component having trans related protons ($^3J=3Hz$). The major alcohol component has $^3J=9-10$ Hz. Reduction of ketone (10) with LiAlD$_4$ gives a similar mixture of isomeric alcohols, the chloromethylene protons appearing now as sharp singlets at the correct calculated positions. In the spectra of these reduction mixtures, the signal due to the ring junction proton H-7 in the minor component - an unmistakable doublet of quartets or triplets - appears at more or less the same relative position with respect to other signals, although there is a general upfield shift of all signals (by ca. 0.1 τ) compared to the ketone (10). Significantly, however, the signal due to the H-7 proton in the major alcohol component is deshielded (ca. 0.3-0.4 τ) and obscured under the downfield signals, consistent with its proximity to the endo-OH group. The ^{1}H nmr evidence for the structure of the ketones (9) and (10) and their mode of formation is therefore entirely self-consistent. Finally, encouraged by some success with computer simulation of the spectrum of ketone (10)-d_3 using an experimental non-iterative programme in conjunction with data from first order analysis, refined chemical shift and coupling constant data are obtained for simulation of the spectrum of ketone (10)-d_3 using the LAOCOON 3 programme.[8] (Tables 1,2). More accurately, the spectrum simulated is the deuterium decoupled spectrum for ketone (10)-d_3 in which line shapes and relative heights are significantly different from the parent 'uncoupled' spectrum. The parameter differences for the two actual spectra are unlikely however to exceed the maximum probable error in the computed readout and are thus useful approximations of the true values for ketone (10)-d_3 and also for ketone (10) itself. Interestingly, the computer simulation reproduces fine structure in the first and third multiplets of the upfield exo-12-H quartet not easily understood in terms of first order analysis.[7]

Apart from the intrinsic scientific interest of the above

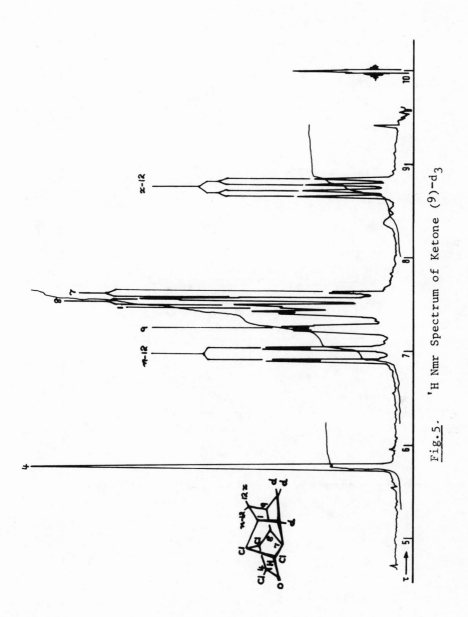

Fig. 5.   ¹H Nmr Spectrum of Ketone (9)-d₃

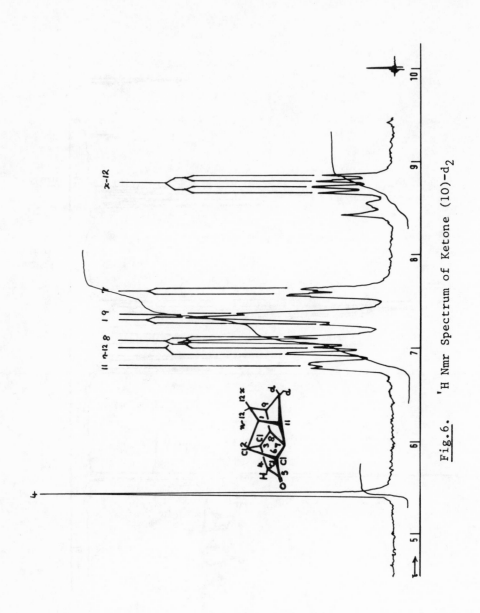

Fig. 6. ^{1}H Nmr Spectrum of Ketone (10)-d_2

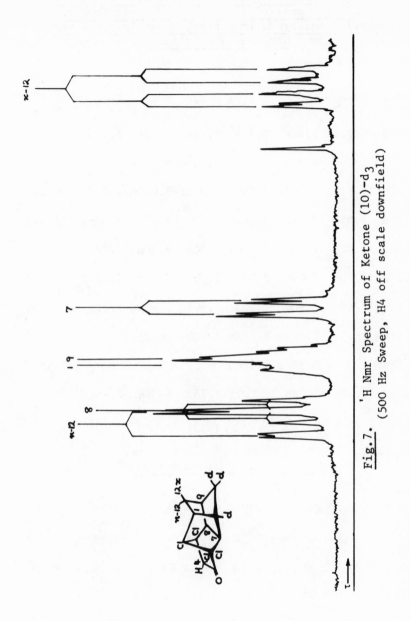

Fig. 7. ¹H Nmr Spectrum of Ketone (10)-d₃
(500 Hz Sweep, H4 off scale downfield)

Table 1: ^{1}H N.m.r. Chemical Shifts of Protons for
Pentacyclododecanones (IX) and (X) and their Deuterium
Labelled Analogues from First Order Analysis and
Computed Simulation

H	IX	IX-d_3	X	X-d_1-a	X-d_1-b	X-d_2	X-d_3	Simulated X-d_3
4	5.75s	5.76s	5.44s	5.44s	5.45s	5.43s	5.43s	−
11	6.82bm	−	6.80bm	6.80bm	6.80bm	6.80bm	−	−
12n	7.00dt	6.97dd	7.00dt	7.00bs	7.00dt	7.00dm	7.01dd	6.991
9	7.27m	7.24m	7.35m	7.34bs	7.37m	7.34m	7.34m	7.340
1	7.40m	7.40m	7.30m	7.30bs	7.31m	7.30m	7.30m	7.302
8	7.50m	7.50m	7.07t	7.07t	7.06m	7.07t	7.07dd	7.067
7	7.60m	7.59dd	7.60dq	7.60dt	7.60dq	7.60dq	7.61dd	7.601
10	8.46dm	−	8.42dm	8.41dm	8.42bs	−	−	−
10′	8.58dm	−	8.58dm	8.58dm	−	−	−	−
12x	8.76q	8.75q	8.76q	−	8.75q	8.76q	8.75q	8.752

Table 2: Coupling Constants

	H8	H1	H12n	H12x	H7
H7	7.11	1.77	0.21	−0.19	−
H9	5.39	2.08	1.85	1.15	−0.15
H12x	0.97	−	−12.74	−	−
H1	−0.20	−	0.1	6.59	−
H8	−	−	0.03	−	−

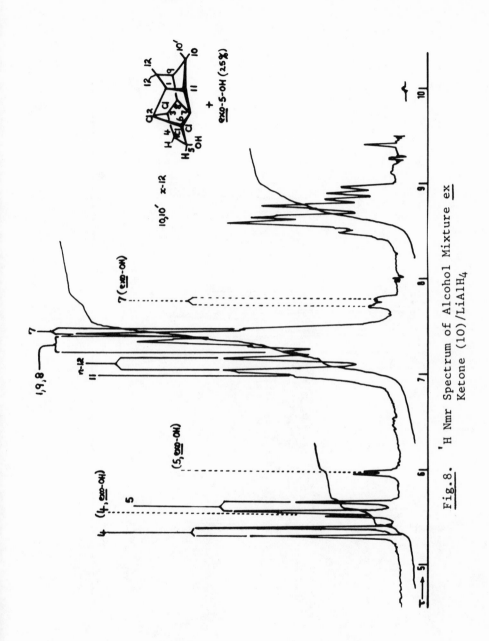

Fig. 8. ¹H Nmr Spectrum of Alcohol Mixture ex Ketone (10)/LiAlH₄

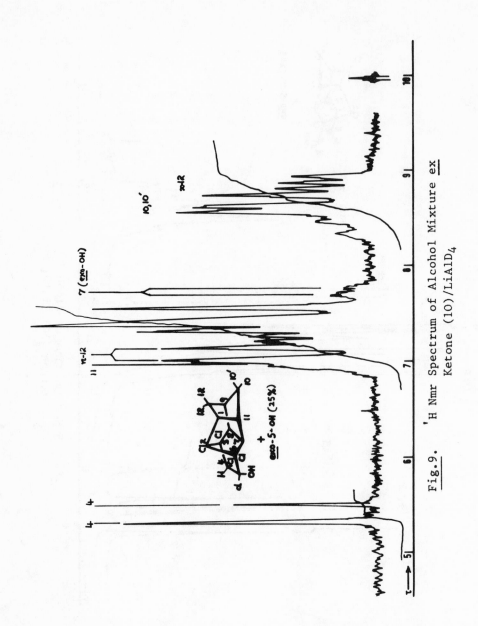

Fig. 9. ¹H Nmr Spectrum of Alcohol Mixture ex Ketone (10)/LiAlD₄

experiments it is possible that these results could be of interest
in connexion with other complex halogenated ring systems,
especially those compounds of unknown structure likely to be
formed as metabolites, and other transformation products of the
polychlorocyclodiene pesticides occurring in the environment.
Most of our results have been obtained with samples of the dienes
(5) and (6) ∿ 500 mg.-1g. but useful ¹H nmr data can be obtained
with much smaller quantities than typically the ∿ 30 mg. samples
of the various ketones (9) and (10) and their labelled analogues
used in this work.

Acknowledgements

We thank the Science Research Council for the award of
Research Studentships for C.H.M.A. and D.J.C., Shell Research Ltd.
for chemical samples, and the University of Bristol and the
Royal Society for the award of travel grants to K.M.

References

1. C.H.M. Adams and K. Mackenzie, J. Chem. Soc., 1969, 480.

2. S.B. Soloway, A.M. Damiana, J.W. Sims, H. Bluestone and
    R.E. Lidov, J. Amer. Chem. Soc., 1960, 82, 5377.

3. C.W. Bird, R.C. Cookson and E. Crundwell, J. Chem. Soc.,
    1961, 4809.

4. L. de Vries and S. Winstein, J. Amer. Chem. Soc., 1960, 82,
    5363.

5. K. Mackenzie, J. Chem. Soc., 1962, 457.

6. S. Winstein, P. Carter, F.A.L. Anet and A.J.R. Bourn,
    J. Amer. Chem. Soc., 1965, 87, 5247.

7. Full details of this and other work described here may
    be found in K. Mackenzie, C.H.M. Adams and D.J. Cawley,
    J. Chem. Soc. Perkin II, 1973 (in press).

8. A.A. Bothner-By and S.M. Castellano in Computer Programmes
    for Chemistry, (Editor D.F. Detar), Benjamin, 1968.

    K. Wiberg, Computer Programmes for Chemists, Benjamin, 1965.

^{13}C NMR STUDIES OF SOME BICYCLO[2.2.2]OCTANONES.[a] CHARACTERIZA-
TION OF THE ISOMERIC 1,5,5,8-TETRAMETHYLBICYCLO[2.2.2]OCTAN-2-ONES

K.R. Stephens, J.B. Stothers and C.T. Tan

Department of Chemistry, University of Western Ontario,

London, Canada   N6A 3K7

ABSTRACT

The ^{13}C nmr spectra of a series of sixteen bicyclo[2.2.2]-
octenones and octanones have been determined to examine the variation of ^{13}C shieldings with methyl substitution. The well-defined
effects arising from 1,4-nonbonded interactions are apparent and
are useful for stereochemical assignments. Evidence is presented
to show that closely neighboring δ carbons experience deshielding
effects, in sharp contrast to the influence of steric crowding on
γ carbons. The ^{13}C shieldings observed for 1,5,8,8-tetramethyl-
bicyclo[2.2.2]oct-5-en-2-one and the isomeric saturated bicyclooctanones obtained therefrom establish their structures.

---

[a] Part 33 in the series ^{13}C NMR Studies; Part 32. J.B. Stothers,
C.T. Tan and K.C. Teo. Can. J. Chem. 00, 0000 (0000).

INTRODUCTION

As an extension of our investigations of the effects of molecular geometry and stereochemistry on ^{13}C shieldings, we have prepared and examined the ^{13}C nmr spectra of a number of methyl substituted bicyclo[2.2.2]octenones and octanones. Because of the closely defined skeletal framework the orientations of substituents in these compounds with respect to the ring carbons are restricted within relatively narrow limits, thereby permitting an assessment of geometrical effects on the observed shieldings for comparison with those found in cyclohexane and norbornane systems, both of which have been extensively studied (1,2,3). From these and numerous other studies (4), it has been established that 1,4 non-bonded interactions between vicinal carbons produce marked shielding changes. These shifts, often called γ effects, are consistently to higher fields (increased shielding) with increased steric crowding and, consequently, are a valuable aid for stereochemical assignments (4c). In the course of the present study, we have synthesized the isomeric 1,5,5,8-tetramethylbicyclo[2.2.2]octan-2-ones, the ^{13}C spectra of which revealed that sterically crowded δ carbons exhibit <u>downfield</u> shifts in striking contrast to the behavior of γ (vicinal) carbons. Although significant downfield shifts of sterically crowded δ-carbons were indicated by the results for some acyclic systems (5), unequivocal evidence from studies of a variety of cyclic systems has only recently been presented (6) to confirm that deshielding effects occur for <u>syn</u>-axial arrangements of δ nuclei presumably via 1,5 nonbonded interactions. We wish to report the results of our examination of some bicyclo[2.2.2]octanones to illustrate the utility of ^{13}C spectroscopy for structural and stereochemical elucidations in these systems and to present additional examples of shielding variations arising from steric crowding of closely neighboring carbons.

Most of the model compounds included in this study were prepared according to the general method described by Cimarusti and Wolinsky (7) as outlined in Scheme 1 and methylation of those products. The 1- and 2-acetoxybicyclo[2.2.2]octanedicarboxylic anhydrides were generated from 2-cyclohexenone, 3-methyl-2-cyclohexenone and isophorone by heating with isopropenyl acetate and maleic anhydride. Bisdecarboxylation of the corresponding 2-oxodiacids with lead tetraacetate led to bicyclo[2.2.2]oct-5-en-2-one (1), 4-methylbicyclo[2.2.2]oct-5-en-2-one (2) and 4,7,7-trimethylbicyclo[2.2.2]oct-5-en-2-one (3), hydrogenation of which furnished the corresponding bicyclo[2.2.2]octan-2-ones 4-6. Methylation of 4 and 5 gave 3-methylbicyclo[2.2.2]octan-2-one (7), 3,4-dimethylbicyclo[2.2.2]octan-2-one (8), 3,3-dimethylbicyclo-[2.2.2]octan-2-one (9) and 3,3,4-trimethylbicyclo[2.2.2]octan-2-one (10). Similarly, <u>exo</u>-3- and <u>endo</u>-3-methylbicyclo[2.2.2]oct-5-en-2-ones (11 and 12) and 3,3-dimethylbicyclo[2.2.2]oct-5-en-2-

SCHEME 1

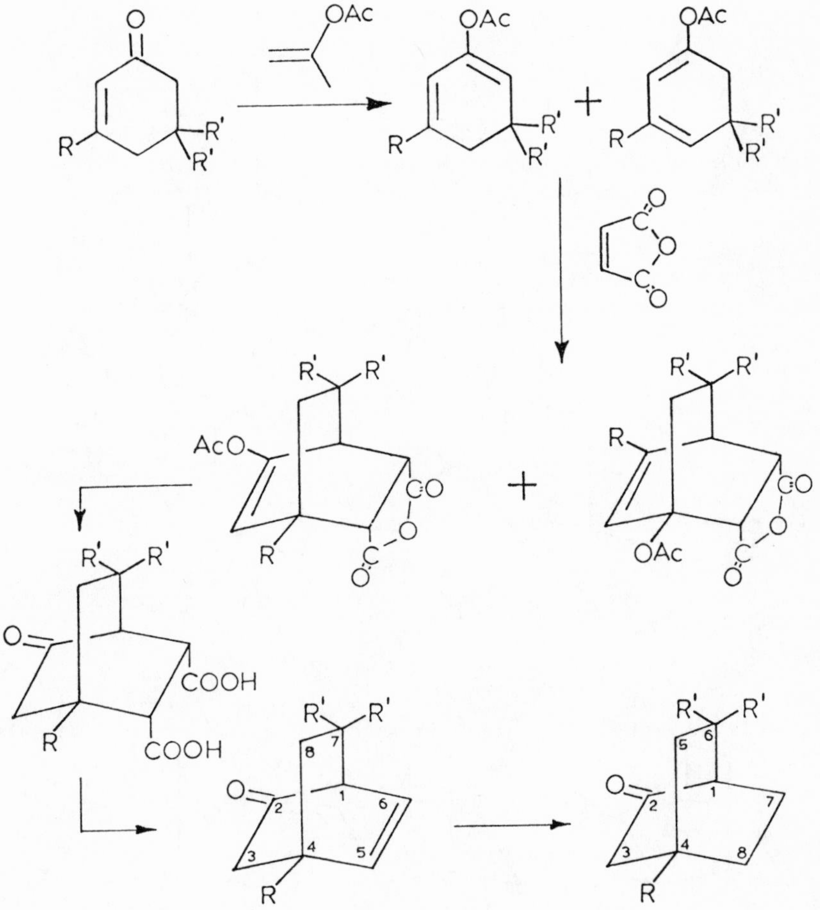

1. R = R' = H
2. R = Me, R' = H
3. R = R' = Me

4. R = R' = H
5. R = Me, R' = H
6. R = R' = Me

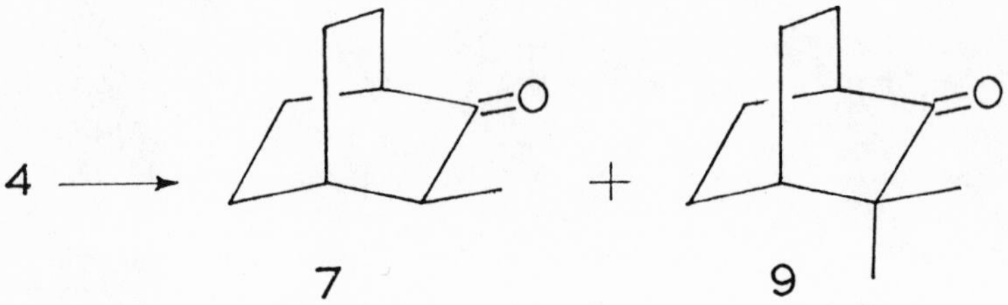

7 9

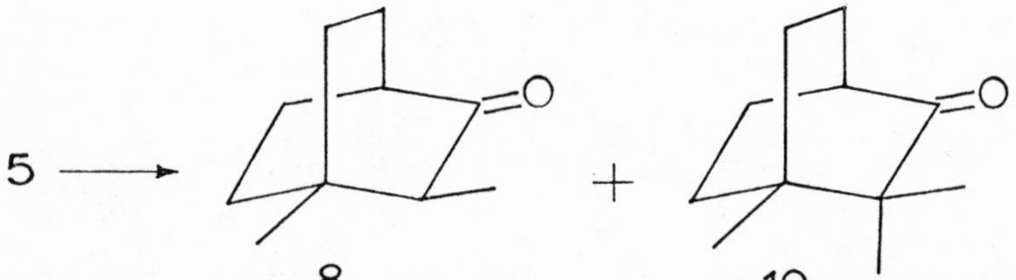

8 10

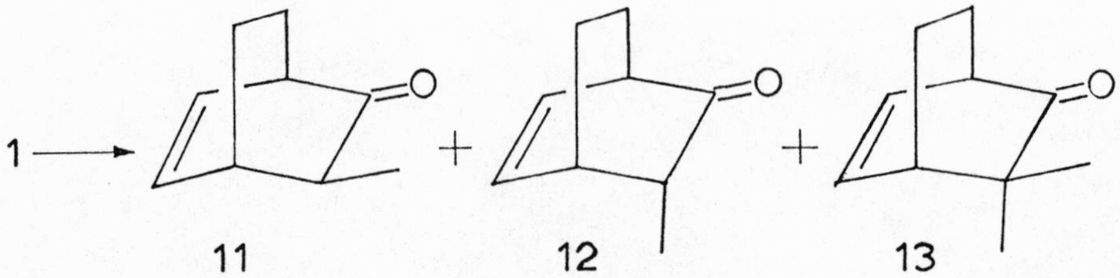

11 12 13

Table 1.   13C Shieldings[a] of some Bicyclo[2.2.2]oct-5-en-2-ones and -octan-2-ones.

Compound	C-1	C-2	C-3	C-4	C-5	C-6	C-7	C-8	Me
Bicyclooctane	24.0	25.2	25.2	24.0	25.2	25.2	25.2	25.2	
Bicyclooctene	29.5	134.1	134.1	29.5	25.8	25.8	25.8	25.8	
1	48.6	212.4	40.4	32.4	136.8	128.3	22.5	24.3	
2	48.7	212.1	47.0	37.4	141.5	127.7	23.8	32.1	24.1
3	61.6	211.6	45.4	37.9	140.6	127.8	35.7	48.8	24.0 (C-4), 29.4, 30.8
4	42.3	216.7	44.6	27.9	24.8	23.4	23.4	24.8	
5	42.1	216.6	50.5	32.6	31.9	23.7	23.7	31.9	27.0
6	54.1	216.3	49.1	33.7	48.8	31.2	19.7	30.8	26.9 (4-Me), 28.8 (anti), 31.9 (syn)
7	42.3	220.1	47.2	33.9	20.2[b]	22.7	24.2	26.1	13.5
8	42.3	219.9	48.7	36.7	26.8[b]	22.8	24.2	34.1	10.4 (3-Me), 24.7 (4-Me)
9	42.7	221.9	45.9	38.5	22.4	23.5	23.5	22.4	23.7
10	42.3	222.2	48.7	36.7	30.0	23.4	23.4	30.0	20.3 (3-Me), 21.3 (4-Me)
11	48.4	215.1	42.5	38.2	136.0	128.1	23.1	18.6	14.4
12	48.4	214.1	44.4	39.1	136.0	127.2	21.9	24.8	17.5
13	48.8	216.7	43.8	44.1	138.6	126.1	21.8	20.7	24.3 (exo), 27.5 (endo)

a   In ppm from internal TMS in CDCl$_3$ solutions.   b   Carbon syn relative to the 3-methyl group.

one (13) were obtained from 1. Another unsaturated ketone, 1,5,8,8-
tetramethylbicyclo[2.2.2]oct~5-en-2-one (14), was prepared as
shown in Scheme 2. The Diels-Alder reaction of α-acetoxyacrylonitrile with cyclohexadiene 15, which was readily prepared from
isophorone (8), gave a single~readily isolable adduct which was
subsequently shown to be 16 as expected since the formation of 17
would be hindered because~of steric interference between the di~~
enophile and the closer of the two geminal methyl groups. Alkaline
hydrolysis of 15 gave 14 which upon hydrogenation over platinum
oxide gave a mixture of~two isomeric saturated bicyclic ketones (18).

RESULTS

The proton noise decoupled ^{13}C spectra of ketones 1-13 were
recorded and the shielding values are listed in Table 1, together
with those of the parent hydrocarbons. The carbonyl and olefinic
signals were readily identified by their low and intermediate
field positions, respectively, and the olefinic assignments for 1
follow from the results for 2. Introduction of a methyl group at
C-4 is expected to deshield C-5 and shield C-6 since this is the
typical behavior in olefinic systems (4a,b). On this basis, the
olefinic assignments for 3 and 11-13 were completed. For the high
field region of these spectra, off~resonance decoupling permitted
the straightforward identification of quaternary, methine, methylene and methyl signals, which can be considered in turn.

The assignments for the two quaternary carbons in 3 and 6
follow from the single quaternary value for 2 and 5, respectively,
since methyl substitution at C-7 should have~little effect at C-4
in each case. The two quaternary centres in 10 were assigned by
comparison with the results for 8 and 9. Of the signals arising
from methine carbons those for C~1 were readily assigned for 1, 2
and 11-13 to the 48.6±0.2 ppm signal because it is uniquely
defined~in 2 and is not expected to be affected by methyl substitution at C~3 or C-4. The characteristic β-effects of the geminal
methyls are clear for C-1 in 3 for which this assignment is unequivocal. Similarly the C-1~signal for 4, 5 and 7-10 must be
that near 42.3 ppm while the C-1 absorption for 6 is~uniquely
defined by off resonance decoupling. The remaining methine signal in the spectra of 1, 4, 9 and 13 must arise from C-4 and in 8
from C-3. Each of 7, 11 and~12 exhibit a pair of methine signals,
in addition to that~for~C-1. The lower field member of each pair
must arise from C-3 since methyl substitution deshields the α and
β carbons and the higher field signal of each pair lies at <40
ppm, i.e. at higher field than C-3 in the parent ketones. The
individual assignments for the methylene carbons are more difficult but a consistent assignment is possible as listed in Table 1
from the values for 4, the analysis of which has been described (9)

SCHEME 2

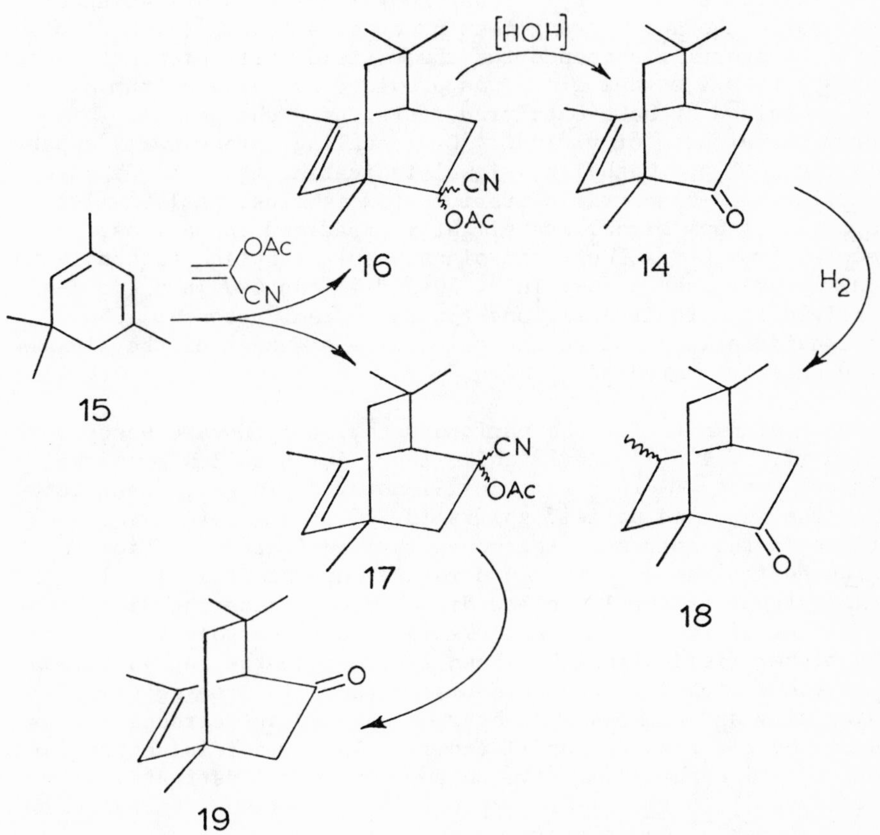

and was confirmed by the data for 5 and 4-3,3 d_2. For the latter compound, obtained by base-catalyzed exchange, the effects of vicinal ^{13}C-D couplings are evident in only one of the remaining methylene signals, that at 24.8 ppm, which must therefore arise from C-5, -8 (10). It may be noted that the expected γ effect of the exo-3-methyl group at C-8 is clearly apparent in 11 and 13 compared to 1. In contrast to the trend in the norbornene derivatives (3) an endo-3-methyl group appreciably deshields the anti γ carbon (C-8). Similar trends attend the variations in the saturated ketones since the 3-methyl group shields the syn γ carbon (C-5 in 7 and 8) while deshielding the anti γ carbon (C-8 in 7 and 8). It follows that the syn arrangement leads to a shielding of 4-5 ppm while the anti arrangement produces a downfield shift of 1-2 ppm. Supporting evidence for the upfield shift of a methylene carbon syn to a γ methyl group was provided by the spectrum of 7-3-d_1, obtained by base-catalyzed deuterium exchange. In this spectrum the effects of vicinal ^{13}C-D coupling were clearly apparent in the 20.2 ppm methylene signal indicating that the dihedral angle relating C-5 and the deuterium atom is close to 180° (10) while the 26.1 ppm signal was slightly broadened and arises, therefore, from C-8. Thus, the signals for methylene carbons γ to geminal methyls such as C-8 in 9, 10 and 13 and C-7 in 6 lie 2-4 ppm upfield from their positions for the parent ketones. The above considerations led to the complete assignment of the signals for the skeletal carbons.

The assignments for the various methyl carbons are unequivocal for 2, 5, 7, 9, 10, 11 and 12 while those for 8 follow from the values for 5 and 7. In 8 the vicinal methyls are gauche and both exhibit the expected upfield shifts (2.7±0.4 ppm) relative to their positions in the corresponding monomethyl derivatives. From the differences for the exo- and endo-methyl carbons in 11 and 12, the methyl assignments for 13 follow directly. Because the difference in shielding of the geminal methyls in 6 is comparable to that in 13 the higher field signal has been assigned to the methyl carbon anti to the carbonyl group. Presumably there is greater steric crowding of γ sp^3-carbons than between sp^2 and sp^3 carbons as evidenced by the results for 11 and 12. The downfield shifts for geminal methyl carbons relative to the monomethyl derivatives (i.e., 9 vs. 7, 10 vs. 8) are typical (4) and arise from the usual β-effect of methyl substitution.

On the basis of the foregoing results, the ^{13}C spectra of 14 and 18 led to the assigned structures; the shieldings are collected in Table 2. The absence of a methyl signal near 24 ppm for the unsaturated ketone indicates that the Diels-Alder adduct (Scheme 2) is 16 rather than 17 since hydrolysis of the latter would yield 19 having a methyl signal equivalent to the bridgehead methyl absorptions of 2 and 3. Furthermore, the geminal methyl signals at 28.6 and 31.1 ppm are consistent with 14 rather than 19 on the basis of

Table 2. ^{13}C Shieldings^a of ketones 14 and 18.

Compound	C-1	C-2	C-3	C-4	C-5	C-6	C-7	C-8	Me
14	49.8	213.1	36.4	49.8	147.1	124.1	46.8	35.0	17.7 (C-1)
									21.9 (C-5)
									28.6 (<u>exo</u>-8-Me)
									31.1 (<u>endo</u>-8-Me)
18 a	44.8	217.2	44.8	45.3	32.5	47.3	38.6	34.4	20.1 (1-Me)
									32.3, 32.8 (5,5-Me$_2$)
									22.6 (8-Me)
18 b	45.0	217.7	35.8	45.4	32.3	47.3	39.3	25.4	20.1 (1-Me)
									29.5 (5,5-Me$_2$)
									20.6 (8-Me)

a/ In ppm from internal TMS in CDCl$_3$ solutions.

the geminal methyl shieldings for 3. From the off resonance decoupled spectrum, the methylene signals were identified at 36.4 and
46.8 ppm. The more shielded of these strongly indicates a methylene carbon experiencing the γ gauche effect of a neighbor methyl
group which is consistent only with 14. The upfield shift of 4 ppm
relative to the C-3 absorption of 1 agrees entirely with that
found for a variety of methylene carbons γ to geminal methyls as
noted above. The sp^3-methine signal at 49.8 ppm is consistent only
with 14 since the corresponding signal in 3 appears at 61.6 ppm.
The appearance of a quaternary signal at 49.8 ppm is also consistent
only with structure 14 since the quaternary bridgehead carbon in
19 would absorb near 38 ppm on the basis of the value for 3. Thus,
these five pieces of evidence establish the unsaturated ketone to
be 14. Since it is known that methyl carbons gauche to, or eclipsed
with, a carbonyl group experience an upfield shift of 6-7 ppm the
17.7 ppm signal was assigned to the bridgehead methyl carbon. Supporting data for this assignment are provided by the methyl shieldings in 20 vs. 21 (2), 22 vs. 23 (2) and 24 vs. 25. The remaining
assignments for 14 were straightforward.

Catalytic hydrogenation of 14 proceeded slowly to give a 4:1
mixture of two saturated bicyclooctanones. From earlier work with
related materials (8), we expected hydrogenation to occur primarily,
if not exclusively, on the side of the olefinic bond away from the
geminal methyls, i.e. syn with respect to the carbonyl group. On
this basis, 18a should represent the major product and 18b the
minor. The ^{13}C data are entirely consistent with expectations.
Only three nonequivalent methyl signals were observed in the spectrum of the minor product at 20.1, 20.6 and 29.5 ppm in the ratio
1:1:2, while four separate methyl signals were found for the major
isomer: 20.1, 22.6, 32.3 and 32.8 ppm. For each ketone, the 20.1
ppm signal was assigned to the bridgehead methyl since the 7 ppm
upfield shift relative to 5 and 6 is expected because of the
vicinal carbonyl bond (see above). Of the two structures, equivalent geminal methyl carbons can only arise in 18b. Since the syn-
8-methyl in 18a lies in close proximity to one of the geminal
methyls the steric perturbation will render the latter nonequivalent. The surprising feature is the fact that these methyl carbons
are deshielded relative to their unhindered counterparts. This
trend is discussed in the next section. To complete the assignments for 18a and 18b, each was subjected to base-catalyzed deuterium exchange and the ^{13}C spectra of the 3,3-dideuterio derivatives
were recorded. From these spectra the signals for C-3, C-4, C-5
and C-8 were unequivocally distinguished by the characteristic
effects of deuterium through 1, 2 and 3 bonds respectively (10).
The C-4 methine signals exhibited geminal isotope shifts of ca.
0.1 ppm while the C-5 and C-8 signals were appreciably broadened
due to vicinal ^{13}C-D coupling. Because deuteration was essentially
complete the C-3 signals were not observed in these spectra. The
relative shieldings of C-3 in 18a and 18b establish that the latter

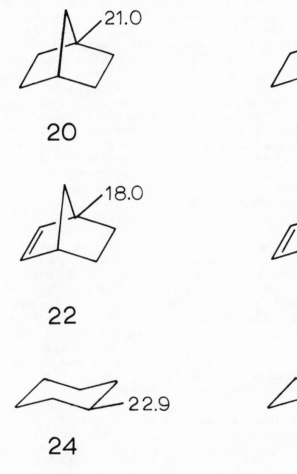

20

21

22

23

24

25

has a γ gauche interaction with the 8-methyl group. The remaining
two methylene signals for each isomer were readily assigned since
the lower field signal must arise from that adjacent to the gem-
dimethyl group.

DISCUSSION

The carbonyl shieldings in this series reveal some interesting trends. Perhaps the most striking of which is the consistent
difference observed between pairs of saturated and unsaturated
ketones with the same substitution pattern, e.g. 4 vs. 1, 5 vs. 2,
etc. In each case, the β, γ-unsaturated ketone exhibits its carbonyl absorption 4.6±0.5 ppm upfield from that in its saturated
counterpart. In an earlier study of related systems (11) this
shift was attributed to homoconjugation but unfortunately the
effect in the [2.2.2] system was missed because of a comparison of
rapid passage data in different solvents from two laboratories.
Nevertheless, it is clear from the present data that the presence
of the double bond produces a significant shielding effect. Since
the olefinic carbons in bicyclo[2.2.2]octene absorb at 134.1 ppm,
the olefinic shieldings in the bicyclooctenones show that the
double bond is polarized toward the carbonyl group in a manner consistent with homoconjugation. While this interpretation of this
behavior in the [2.2.1] system has been criticized (26), the fact
that the methylene carbon shieldings in bicyclo[2.2.2]octene and
bicyclo[2.2.2]octane differ by only 0.6 ppm indicates that the
double bond itself has no profound effect as it may in the [2.2.1]
system. Homoconjugative interaction between the double bond and
carbonyl group may be expected to reduce the electron withdrawing
influence of the carbonyl at C-3 and the relatively higher shielding of C-3 in the unsaturated ketone of each pair is consistent
with this notion. As expected methyl substitution at C-3 tends to
deshield the carbonyl carbon significantly while more remote methyl
substitution has little effect. It is interesting that a 1-methyl
group is less deshielding than a 3-methyl to judge from the results for 14 and 18 relative to 1 and 4, respectively. Perhaps a
steric interaction between the 1-methyl hydrogens and the carbonyl
oxygen tends to diminish the deshielding effect since the dihedral
angle between the C-Me and C=O bonds is appreciably smaller than
that for a 3-methyl group. The higher field position of the 1-
methyl carbon relative to the other methyls is consistent with this
suggestion.

Several instances of upfield shifts arising from 1,4 nonbonded
interaction between neighboring γ nuclei have been cited in the
discussion of individual assignments in the previous section. A
comparison of some of these with other systems seems warranted.
The shielding differences between exo- and endo-methyl groups in
11-13 are appreciably larger than those in the corresponding nor-

bornenones (3). In the present system, the exo-methyl lies closer
to C-8 than an exo-methyl relative to C-7 in the norbornenones
which will account for the observations. In a similar manner, the
larger difference between geminal methyls in 6 relative to that for
camphor or 7,7-dimethylnorcamphor (3) seems attributable to the
closer approach of the anti-methyl to C-7 in 6 than the corresponding interaction of the anti-methyl with C-5, C-6 in the camphor
series. It should be noted that a specific assignment of the
geminal methyls in 3 has not been indicated in Table 1. By comparison with the results for 6, however, it may be suggested that
the lower field signal of the geminal methyl pair in 3 arises from
the exo-6-methyl carbon. The geminal assignments for 6 were based
on the fact that the anti-methyl suffers the greater steric crowding with the methylene carbon than the syn-methyl with the carbonyl
group. For both 3 and 6, the least shielded methyl signal was most
strongly affected by the addition of the shift reagent, Eu(fod)$_3$,
supporting the above assignments.

As discussed in the preceding section, the geminal methyl signals in the 18 isomers provided a clear distinction between the two

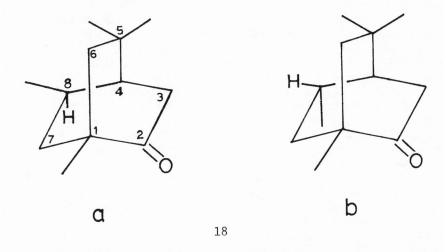

18

saturated ketones since equivalent gem-dimethyl carbons are only
expected for the minor isomer 18b. The other evidence substantiat-
ing this conclusion has been discussed above. Initially it was
surprising to us that the geminal methyl carbons are deshielded in
18a in which there must be steric interference with the 8-methyl
group because it is well-established that steric crowding causes
upfield shifts between γ nuclei (4). Steric crowding of δ nuclei,
however, appears to have the opposite effect. Moreover, both gem-
dimethyl carbons are appreciably deshielded in 18a, as is the 8-
methyl carbon, relative to the corresponding carbons in 18b. The

relative orientation of the neighboring methyls in 18a is close to syn-axial with the flexibility of the ring system permitting some distortion to maximize the separation as depicted in 26. Concur-

26

rent with this work, we were also examining a variety of bicyclic alcohols and found similar trends. In each case having a syn-axial arrangement of hydroxyl and methyl groups the methyl carbon exhibited a downfield shift relative to its shielding for a less hindered environment (6). It has been recognized that, in general, hydroxyl and methyl shielding effects are remarkably similar (12) and the results for the methyl carbons in 18a provide further evidence of this similarity. At present, work is in progress to study the shielding variations of δ methyl groups in sterically crowded environments through the preparation and examination of additional model systems. The presently available data, however, are sufficient to show that interpretations of small shielding variations in complex systems must be approached cautiously since steric interactions can apparently lead to both upfield and down-field shifts. On the basis of the upfield shifts consistently observed for γ nuclei, there has been a developing tendency to associate steric crowding in general with increased shielding. Un-fortunately, it seems that this generalization is somewhat pre-mature.

EXPERIMENTAL

Materials. As noted earlier, ketones 1, 3, 4 and 6 were pre-pared by published procedures (7) and the same sequence was em-ployed for 2 and 5 as described below. Methylation of 1, 4 and 5 using the sodamide method of Corey et al. (13) furnished mixtures of the mono- and dimethyl derivatives 7-10 and 11-13. In each case the mixtures were separated by gas chromatography using a 20%

SE30 column. Proton spectra served to identify the individual components. Of these derivatives, 7 has been described (14) and the physical constants agreed well with those reported.

4-Methylbicyclo[2.2.2]oct-5-en-2-one (2). A mixture of maleic anhydride (33.0g), 3-methylcyclohex-2-enone~(30.0g) and isopropenyl acetate (50.0g) containing a trace of p-toluenesulphonic acid was heated under reflux for 72 h. After cooling, the crystalline product was separated into two components by fractional crystallization from ethyl acetate. The higher melting component, m. 145.0-145.5°, 20% yield, was shown to be endo-1-acetoxy-3-methylbicyclo-[2.2.2]oct-2-ene-5,6-dicarboxylic anhydride ν_{max} 1850, 1780, 1735 (C=O), 1650 cm^{-1} (C=C).

Anal. Calcd. for $C_{13}H_{14}O_5$: C, 62.39; H, 5.64.
Found: C, 62.35; H, 5.70.

The major adduct, endo-1-methyl-3-acetoxybicyclo[2.2.2]oct-2-ene-5,6-dicarboxylic anhydride was isolated in 44% yield and identified by its proton and infrared spectra: ν_{max} 1865, 1840, 1770 (C=O), 1655 cm^{-1} (C=C); m. 142-144°. Hydrolysis by heating in water furnished the corresponding ketodicarboxylic acid which was converted to dimethyl endo-4-methylbicyclo[2.2.2]octan-2-one-5,6-dicarboxylate by treatment with diazomethane, m. 70-72°.

Anal. Calcd. for $C_{13}H_{18}O_5$: C, 61.40; H, 7.31.
Found: C, 61.45; H, 7.31.

Bisdecarboxylation of the diacid according to the published method (7) gave a 50% yield of 4-methylbicyclo[2.2.2]oct-5-en-2-one (2), b. 85°/7 mm; oxime, m. 142.5-143.5°.

Anal. Calcd. for $C_9H_{12}O$: C, 79.37; H, 8.88.
Found: C, 79.01; H, 9.06.

4-Methylbicyclo[2.2.2]octan-2-one (5). Catalytic hydrogenation of 2 in ethyl acetate over platinum~oxide gave 5, b. 85°/7 mm, n_D^{25} 1.4753.

Anal. Calcd. for $C_9H_{14}O$: C, 78.21; H, 10.21.
Found: C, 77.70; H, 10.03.

The oxime of 5 melted at 130-131°.

Anal. Calcd. for $C_9H_{15}NO$: C, 70.55; H, 9.87; N, 9.14.
Found: C, 70.69; H, 9.86; N, 9.04.

1,5,8,8-Tetramethylbicyclo[2.2.2]oct-5-en-2-one (14). The Diels Alder reaction of 18.5g α-acetoxyacrylonitrile (15) with 15 (27.3g) was carried out in refluxing t-butylbenzene (20 ml) for~8 h.

The crude product was isolated in 20% yield by steam distillation and extraction of the distillate with methylene chloride. Recrystallization from methanol gave pure 16, m. 105.5-106°; ν_{max} 2270 (CN), 1745 (C=O), 1655 cm^{-1} (C=C).

Anal. Calcd. for $C_{15}H_{21}NO_2$: C, 72.84; H, 8.53; N, 5.66.
Found: C, 72.92; H, 8.56; N, 5.85.

A solution of 16 (4.8g) in methanol (30 ml) was added to 3M NaOH (30 ml) and refluxed for 1 h. Ether extraction afforded a light yellow oil which was purified by vacuum sublimation to yield pure 14, m. 35.0-36.5°, ν_{max} 1720 (C=O), 1605 cm^{-1} (C=C).

Anal. Calcd. for $C_{12}H_{18}O$: C, 80.84; H, 10.17.
Found: C, 80.88; H, 10.05.

1,5,5,8-Tetramethylbicyclo[2.2.2]octan-2-one (18). To a solution of 14 (1.7g) in glacial acetic acid (15 ml) was added platinum oxide (200 mg) and the mixture shaken for 42 h under hydrogen pressure of 55 psi. Gas chromatography on a 20% SE30 column showed the presence of two components in a 4:1 ratio. Preparative gas chromatography was used to obtain suitable samples of each.

The major fraction 18a had the following properties: m. 34.5-35.0°; ν_{max} 1710 cm^{-1} (C=O).

Anal. Calcd. for $C_{12}H_{20}O$: C, 79.94; H, 11.18.
Found: C, 79.70; H, 11.30.

The minor fraction was distilled to obtain 18b: b. 80°/16 mm; ν_{max} 1712 cm^{-1} (C=O).

Anal. Calcd. for $C_{12}H_{20}O$: C, 79.94, H, 11.18.
Found: C, 79.72; H, 11.07.

Methylation of 4. Following the method described by Corey et al. (13), the sodio derivative of 4 was generated using sodium amide in ether and methylated with methyl iodide. After the usual isolation procedure, the product was analyzed by gas chromatography and found to contain 21% 4, 24% 7 and 55% 9. The individual components were separated by preparative gas chromatography at 130° using a SE30 column. Proton and infrared spectra were used to identify 7 and 9.

3,3-Dimethylbicyclo[2.2.2]octan-2-one (9): ν_{max} 1705 cm^{-1}; m/e, 152.

Anal. Calcd. for $C_{10}H_{16}O$: C, 78.89; H, 10.59.
Found: C, 79.05; H, 10.81.

Methylation of 5. In a similar manner, methylation of 5 gave
a mixture containing~31% 5, 40% 8 and 29% 10, separated by preparative gas chromatography on SE30 at 117°~. Proton and infrared
spectra were employed to identify each of the methylated products
and precise molecular ion masses were determined:
    8, m. 36.0-36.5°; ν_{max} 1707 cm^{-1}; molecular weight calculated
for $C_9H_{16}O$: 152.1200. Found: 152.1211.
    10, m. 90-91°; ν_{max} 1707 cm^{-1}; molecular weight calculated for
$C_{11}H_{18}O$: 166.1357. Found: 166.1355.

Methylation of 1. Using the same procedure, methylation of 1
gave a mixture consisting of 20% 1, 54% (11 and 12) and 26% 13. ~
Preparative gas chromatography on~SE30 at~100° furnished samples
of the methylated products but failed to separate the monomethyl
derivatives. ^{13}C nmr showed the ratio of 11:12 to be ca. 1:2.
Proton and infrared spectra confirmed the structures for each.
    11 and 12, ν_{max} 1717 (C=O), 1615 cm^{-1}(C=C). Anal. Calcd. for
$C_9H_{12}O$: C, 79.37; H, 8.88. Found: C, 79.50; H, 9.01.
    13, m. 89-90°; ν_{max} 1710 (C=O), 1610 cm^{-1}(C=C). Molecular
weight~calculated for $C_{10}H_{14}O$: 150.1044. Found: 150.1039.

^{13}C Spectra. A Varian XL-100-15 spectrometer operating in
the Fourier transform mode was employed for the ^{13}C spectra. All
samples were examined in $CDCl_3$ solutions (10-20%) in 5 mm spinning
tubes. The positions of the noise-decoupled peaks were measured
relative to internal TMS with a precision of ±1 Hz (at 25.2 MHz)
using 2 KHz sweep "windows". Off resonance decoupled spectra were
obtained by single frequency proton irradiation approximately 1 KHz
off resonance. For individual samples 500-3000 transients were
collected at a repetition rate of 1 sec.

ACKNOWLEDGEMENTS

    We wish to thank the National Research Council of Canada for
financial support of this project. We are grateful to Miss J.L.
Gough and Mrs. C. Ferriera for some technical assistance.

REFERENCES

1. D.K. Dalling and D.M. Grant. J. Am. Chem. Soc. 89, 6612 (1967);
    94, 5318 (1972).

2. (a) J.B. Grutzner, M. Jautelat, J.B. Dence, R.A. Smith and
    J.D. Roberts. J. Am. Chem. Soc. 92, 7107 (1970); (b) E. Lippmaa,
    T. Pehk, J. Paasivirta, N. Belikova and A. Plate. Org. Mag.
    Res. 2, 581 (1970).

3. J.B. Stothers, C.T. Tan and K.C. Teo. Can. J. Chem. $\underset{\sim\sim}{00}$, 0000 (0000).

4. For general discussions see (a) G.C. Levy and G.L. Nelson. ^{13}C Nuclear Magnetic Resonance for Organic Chemists. Wiley-Interscience. New York, 1972; (b) J.B. Stothers. Carbon-13 NMR Spectroscopy. Academic Press. New York, 1972; (c) N.K. Wilson and J.B. Stothers in Topics in Stereochemistry. Vol. 8. edited by E.L. Eliel and N.L. Allinger. Wiley-Interscience, New York. 1973. Chap. 1.

5. J.I. Kroschwitz, M. Winokur, H.J. Reich and J.D. Roberts. J. Am. Chem. Soc. $\underset{\sim\sim}{91}$, 5927 (1969).

6. S.H. Grover, J.P. Guthrie, J.B. Stothers and C.T. Tan. J. Mag. Res. $\underset{\sim\sim}{00}$, 0000 (0000).

7. C.M. Cimarusti and J. Wolinsky. J. Am. Chem. Soc. $\underset{\sim\sim}{90}$, 113 (1968).

8. Gurudata and J.B. Stothers. Can. J. Chem. $\underset{\sim\sim}{47}$, 3515 (1969).

9. G. van Binst and D. Tourwe. Org. Mag. Res. $\underset{\sim}{4}$, 625 (1972).

10. J.B. Stothers, C.T. Tan., A. Nickon, F. Huang, R. Sridhar and R. Weglein. J. Am. Chem. Soc. $\underset{\sim\sim}{94}$, 8581 (1972).

11. Gurudata and J.B. Stothers. Can. J. Chem. $\underset{\sim\sim}{47}$, 3601 91969).

12. J.D. Roberts, F.J. Weigert, J.I. Kroschwitz and H.J. Reich. J. Am. Chem. Soc. $\underset{\sim\sim}{92}$, 1338 (1970).

13. E.J. Corey, R. Hartmann and R.A. Vatakencherry. J. Am. Chem. Soc. $\underset{\sim\sim}{84}$, 2611 (1962).

14. A. Orahovats, M. Tichy and J. Sicher. Coll. Czech. Chem. Comm. $\underset{\sim\sim}{35}$, 838 (1970).

15. R.M. Nowak. J. Org. Chem. $\underset{\sim\sim}{28}$, 1182 (1963).

CARBON–13 AND PROTON MAGNETIC RESONANCE STUDIES OF CHLORINATED

BIPHENYLS

Nancy K. Wilson and Marshall Anderson

National Institute of Environmental Health Sciences

P. O. Box 12233, Research Triangle Park, N. C. 27709

Abstract: ^{13}C and ^1H nuclear magnetic resonance spectra were obtained for ten symmetric chlorinated biphenyls and for 4,4'-disubstituted biphenyls with NH_2, CH_3O, CH_3, Cl, F and NO_2 substituents. The ^1H shieldings are shown to correlate with π electron densities calculated by the CNDO/2 method, if corrections are made for ring current effects from the second ring. For freely rotating biphenyls, additive substituent parameters obtained from benzene data predict the ^{13}C shieldings with reasonable precision. Steric hindrance to rotation by substituents at the 2,6,2' and/or 6' positions causes deviations from the substituent effects predicted by additivity; these deviations in δ_C are especially pronounced for the carbons at positions 1 and 1'. The total charge correlation with δ_C in substituted benzenes is also valid in non-hindered substituted biphenyls. Small differences in δ_C between chlorinated biphenyl isomers are best related to differences in the calculated σ electron densities.

Since the discovery a few years ago that polychlorinated biphenyls (PCBs) are not only widespread but also remarkably persistent in the natural environment, investigations centering about their identification in complex mixtures and their interactions with biological systems have been numerous.[1] Most of the analytical studies have concerned characterization of the components of mixtures of PCB isomers, of which there are more than 200 possible, having various degrees of chlorination. Emphasis has been on the use of chromatographic and mass spectroscopic techniques.[2] Similarly, the majority of the inquiries into the possible influences of PCBs on living organisms has utilized commercial chlorinated biphenyl mixtures, rather than pure separate isomers.[3]

197

Several programs at NIEHS have recently been started, to investigate the properties and biological effects of individual chlorinated biphenyls. A number of PCBs were synthesized, to serve as standards for identification and to be used in the biological studies. Nuclear magnetic resonance, as an analytical tool sensitive to both molecular structure and molecular interactions, is useful for discrimination between the many PCB isomers, especially in conjunction with GLC and MS data. Because nmr parameters are markedly dependent on such molecular characteristics as electronic charge distribution, the observation of trends in these parameters might enhance understanding of the modes of interaction of PCBs with biological systems. Thus, this study had two aims: to determine nmr shieldings and coupling constants for some pure chlorinated biphenyls, and to investigate the relationships of these parameters to charge densities in the chlorinated biphenyl molecules in order to allow more meaningful examinations of interactions of PCBs with biologically significant compounds.

Proton nmr shieldings and approximate coupling constants obtained at 220 MHz for a large number of PCBs fractionated from a commercial mixture have been reported by Welti and Sissons.[4] Bartle has measured these at 60 MHz for two other PCBs.[5] Detailed analyses of the ^{1}H spectra of symmetrically substituted dihalobiphenyls have been reported by Tarpley and Goldstein.[6] There are also several earlier studies of proton shieldings and substituent effects on these in halobiphenyls.[7] Fewer ^{13}C nmr studies of biphenyls have been reported. These include one of alkylbiphenyls[8] and several of biphenyl itself[9-11] and 3- and 4-bromobiphenyl.[12]

Numerous quantum mechanical calculations have been made for biphenyl,[13,14] most with the goal of predicting its equilibrium conformation and barrier to internal rotation. A few extensions to alkyl-substituted biphenyls have been made.[15,16] Among these are the attempt, which was moderately successful, by Hasegawa et al.[8] to predict the ^{13}C shieldings in seventeen alkylated biphenyls using Hückel molecular orbital calculations and the Karplus-Pople equation.[17] Calculations on several fluorinated biphenyls have been carried out by Farbrot and Skancke,[18] and on various ortho bromine- and iodine-substituted biphenyls by Pedersen[19] and earlier by Howlett.[20] Applications of semi-empirical MO techniques to halogenated biphenyls have not as yet been extensive, and this is not surprising, since most of these semi-empirical methods have not been parameterized to include second row elements, or if they have, as is true for CNDO/2[21] and recently for INDO,[22] the parameterizations are still preliminary.[23]

Although the CNDO/2 method has had varying degrees of success in predictions of molecular conformations, and has occasionally led to predictions inconsistent with experimentally determined molecular

geometries[13] and potential energy curves,[24,25] it has been reasonably good at prediction of dipole moments and molecular charge
distributions.[26] Several studies have shown that charge densities
calculated by the CNDO/2 method correlate well with experimental
trends in nmr data.[27-29] Thus in this study, the CNDO/2[30a] technique, employing experimental geometries if these are known and otherwise the Pople bond lengths and bond angles,[26] was used to predict
charge distributions in some biphenyls which can be expected to be
freely rotating at ambient temperatures (i.e., biphenyls with
hydrogen at the ortho positions). These predictions have been
compared with the nmr shieldings for both ^{1}H and [13]C.

    Proton Spectra: Nmr data were first obtained for several non-
chlorinated 4,4'-disubstituted biphenyls, to be used in investigation
of correlations of the nmr parameters with calculated electron
densities. The proton shieldings and coupling constants obtained
by computer analysis of the spectra of these 4,4'-disubstituted
biphenyls, using the program LAOCN3,[30b] are given in Table 1.
Since for these AA'BB' spectra, interchanging the proton resonance
frequencies assigned to the H-2,6 and the H-3,5 pairs will not
affect the appearance of the spectra, except for 4,4'-difluorobi-
phenyl, which is of type AA'BB'X, the assignments are based on the
approximate additive parameters of Tarpley and Goldstein,[6] known
substituent effects on proton resonances in monosubstituted benzenes[31] and the relative calculated π electron densities at these
positions.

    Attention was next directed to symmetric chlorinated biphenyls,
since interpretation of the nmr spectra for these is much simpler
than for the asymmetric isomers. Data for ten symmetric di-, tetra-
and hexa-chlorobiphenyls are reported here.

    The proton nmr spectra of most of the PCBs were simple first-
order spin patterns, or second-order patterns such as A_2B, for which
shieldings and coupling constants could be obtained directly from
the spectra or calculated with little additional effort. The proton
nmr parameters obtained are listed in Tables 2 and 3.

    Several characteristics of the proton spectra for the PCBs may
be pointed out. The coupling constants given in Table 2 are within
the range of those in other substituted aromatic compounds;[31] for
the PCBs, $J_o \simeq 8$ Hz, $J_m \simeq 2$ Hz and $J_p \simeq 0.4$ Hz. The effects on the
coupling constants of varying the chlorine substitution are small.
A chlorine substituent has the expected deshielding effect on the
protons in adjacent (ortho) positions, which is evident in the data
in Table 3 for all the PCBs. An example is the deshielding, resulting in a larger value of δ, of H-3 relative to H-5 in 2,2'-dichloro-
biphenyl. Other influences on the ^{1}H shieldings are not easily
separated from one another, although because of their $1/r^3$ dependence,

Table 1

Proton Shieldings and Coupling Constants for 4,4'-disubstituted Biphenyls[a]

| Subst. | δ, ppm | | J, Hz | | | | | | |
	H-2,6	H-3,5	23,56	25,36	26	35	2F	3F	2F'
CH_3	7.4648	7.2247	7.80	0.12	1.89	1.89			
CH_3O	7.4718	6.9530	8.56	0.11	2.37	2.37			
NO_2	7.7863	8.3636	8.69	0.14	2.11	2.11			
NH_2	7.3370	6.7238	8.49	0.11	2.20	2.20			
Cl	7.4019	7.4662	8.36	0.40	2.34	2.34			
F	7.4807	7.1248	8.71	0.09	2.60	2.61	5.24	8.70	0.19
F[b]	7.0393	6.8173	8.58	0.37	2.75	2.56	5.21	8.56	0.16

[a] At 28°C; 2.5 mole% in chloroform-d, except as noted.

[b] 1-2 mole% in benzene-d_6, from Reference 6.

Table 2

Proton Coupling Constants for Symmetric Chlorinated Biphenyls[a]

Substitution	23	24	25	26	34	35	36	45	46	56
					J, Hz					
2,2'[b]			0.46		8.07	1.29	0.42	7.53	1.68	7.66
3,3'[b]		2.09		1.78				8.00	1.02	7.91
4,4'	8.36		0.40	2.34		2.34	0.40			8.36
2,4,2',4'						1.94	0.44			8.25
2,6,2',6'					8.12			8.12		
2,4,6,2',4',6'										
2,4,5,2',4',5'							≤0.2			
3,4,3',4'			0.45	2.10						8.35
3,5,3',5'										
3,4,5,3',4',5'										

[a] At 28°C; 2.5 mole% in chloroform-d, except as noted.

[b] Approximately 3.5 mole% in benzene-d$_6$.

Table 3

Proton Shieldings for Symmetric Chlorinated Biphenyls[a]

Substitution	H-2	H-3	H-4	H-5	H-6
2,2'[b]		7.2521	6.8694	6.9071	7.0053
3,3'[b]	7.2698		7.0644	6.8456	6.9222
4,4'	7.4019	7.4662		7.4662	7.4019
2,4,2',4'		7.5096		7.3188	7.1834
2,6,2',6'		7.4482	7.3299	7.4482	
2,4,6,2',4',6'		7.4677		7.4677	
2,4,5,2',4',5'		7.6198			7.3641
3,4,3',4'	7.6184			7.5142	7.3581
3,5,3',5'	7.5890		7.5890		7.5890
3,4,5,3',4',5'	7.5360				7.5360

δ, ppm

[a]At 28°C; 2.5 mole% in chloroform-d, except as noted.

[b]Approximately 3.5 mole% in benzene-d6.

field effects and diamagnetic anisotropy cannot be expected to
effect general trends in the shieldings except for effects on
directly bonded or closely neighboring protons. In biphenyls, two
types of ring current effects must be considered. The first of
these results from π electron circulation in the ring to which a
proton is attached. This effect is similar to that in benzenes,
that is, ring current increases lead to deshielding of the protons,
and the magnitude of the deshielding depends on both the nature and
degree of substitution, with increased substitution leading to decreased ring current and hence increased shielding.[6,32-34]
The second type of effect is due to the presence of the other
aromatic ring. This is also deshielding, and for a given substitution pattern in the second ring, the magnitude of this effect
depends on the dihedral angle between the rings. As the inter-ring
angle is increased, the effect of ring current in the second ring
on ^{1}H shieldings in the first ring becomes smaller.[33] The influence
of the ring current in the second ring on the resonances of protons
on the first ring can be nicely demonstrated if the dependence of
δ_H on π charge density in benzenes is compared with that for biphenyls. If the biphenyl data are corrected for the ring current
contribution from the second ring, both plots have similar linear
correlations. Additionally, protons in biphenyls with ortho
substituents are shielded relative to non-ortho-substituted biphenyls
due to the larger torsional angle.

One of the important contributors to proton shieldings is known
to be the π-electron density at the position of interest.[31] Proton
shieldings in monosubstituted benzenes have been thoroughly investigated in this respect;[28,32,35] the para shieldings were found to
correlate best with π charge densities, at the ortho position both
π and total charge densities were major determinants of the shieldings, and at the meta position, ring current effects appeared to be
most significant in determining the order of the meta shieldings.[28]

Similar effects in the biphenyls might be expected, with due
allowance for the influence of the second ring. For the series of
4,4'-disubstituted biphenyls, in which the variation of substituents would be expected to have its primary effect on the π electron
density within the ring due to variations in the resonance interaction, a linear correlation of ^{1}H shieldings with total charge was
found, but a better correlation was obtained for the π charge. [All
CNDO charge densities used here are calculated for a dihedral angle
of 42° between the rings, which corresponds to that measured for
biphenyl itself by electron diffraction.] This plot is shown in
Figure 1. For this plot, the proton shieldings at the 2 and 3
positions were corrected for the difference in the ring current
effect of the second ring using values determined by Mayo and
Goldstein[33] for biphenyl. If only the shieldings of protons at
positions 3 and 5 are considered, the fit is very good, with the

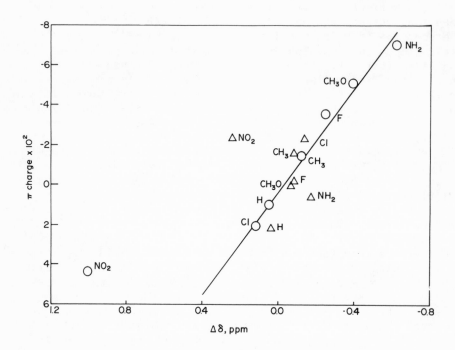

Fig. 1. π charge density vs. ^{1}H shielding relative to
benzene for 4,4'-disubstituted biphenyls. The data were
obtained at 28°C for 2.5 mole% solutions in chloroform-d.
δ_H values were corrected for the ring current effects
from the second ring.[33]   O H-3,5;   △ H-2,6.

exception of the points for 4,4'-dinitrobiphenyl. Points for protons 2 and 6 are more scattered, in agreement with results for substituted benzenes. Since the nitro substituent is known to alter the geometry of the ring considerably,[29] and this alteration was not taken into account in the calculations, the discrepancy for the dinitrobiphenyl points is probably not significant. The degree to which the amino-substituted biphenyl fits this correlation also depends markedly on the geometry of the substituent. In these calculations some distortion of the NH_2 group toward a planar geometry was included.

For the PCBs, if again discussion is restricted to those isomers which are essentially freely rotating at ambient temperature, a linear relationship is found between the π charge density at a given carbon and the shielding of the attached proton. This is illustrated in Figure 2. The line shown in this figure is the same as that in the preceding figure, and the points correspond to protons at positions 2,6 and 3,5 in biphenyl itself and chlorinated biphenyls lacking substituents _ortho_ to the second ring. Again a correction for the difference in the ring current effect of the second ring on H-2 and H-3 has been applied. Although these points are clustered about the same line as in Figure 1, the total range of the variables is not large, and reflects the relatively small effect of varying chlorine substitution on the π electron density in the aromatic ring.

Carbon-13 Spectra: The ^{13}C nmr spectra obtained in this study are all Fourier-transformed spectra for samples at natural abundance, and represent averages of 500 to about 10,000 transients. Since pseudo-random proton noise decoupling was used, all peaks in a given spectrum are singlets, as is shown for 3,3'-dichlorobiphenyl in Figure 3. Assignments were nevertheless straightforward, since shielding additivity relationships for ^{13}C spectra are more reliable than for proton spectra, particularly if intercomparisons are made for series of similar compounds.[36-38] For most of the PCBs, and for all of the 4,4'-disubstituted biphenyls, additive substituent parameters determined from ^{13}C spectra of substituted benzenes predicted the approximate shieldings in the biphenyls.[28,38] Further assistance was available in the form of ^{13}C shielding data for the mono- and di-halobenzenes in papers by Tarpley and Goldstein.[39] Guidance in assignments was also provided by the intensity relationships of the various peaks in the spectra. Thus quaternary carbons such as C-1 and C-1', the carbons involved in the inter-ring bond, have much lower intensities than do an equal number of protonated carbons under the experimental conditions employed, because they lack the strong proton-carbon dipolar spin-lattice relaxation and hence lack the nuclear Overhauser enhancement operative for carbons bearing protons. Additionally, the quaternary carbon spin-lattice relaxation times are long and thus their nmr absorption intensities

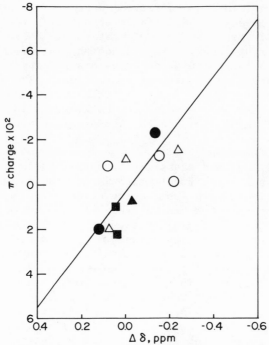

Fig. 2.   π charge density vs. ^{1}H shielding relative to benzene for freely rotating chlorinated biphenyls. The data were obtained at 28°C for 2.5 mole% solutions in chloroform-d̲.   δ$_H$ values were corrected for ring current effects from the second ring.[33]
O 3,3'; ● 4,4'; △ 3,4,3',4'; ▲ 3,5,3',5'; ■ nil

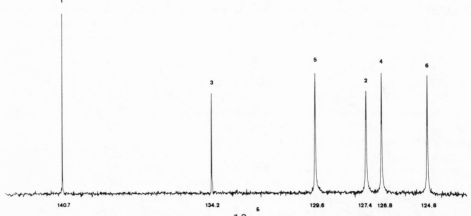

Fig. 3.   Proton noise – decoupled ^{13}C spectrum of 3,3'-dichlorobi-phenyl. The data were obtained at 40°C for an ≈0.6 M̲ solution in 1,2-dibromoethane-d̲4. The spectrum represents an average of 1870 transients, obtained with a 60 μs pulse and an acquisition time of 4.0 s on a Varian XL-100 spectrometer.

are small relative to those of protonated carbons with the short
pulse separations ($\simeq$2-4 sec) used in the experiments. The protonated carbons in the biphenyls, however, have __approximately__ the same
spin-lattice relaxation times and Overhauser enhancements, so that
the relative intensities of their resonances approximately reflect
the relative numbers of ^{13}C nuclei of each type.

The ^{13}C shieldings measured for the series of 4,4'-disubstit-
uted biphenyls are listed in Table 4. The range of shieldings is
approximately 48 ppm, with the largest variation shown by C-4, the
substituted carbon. The C-1 resonance is consistently downfield
from those of the protonated carbons, although its position relative
to the C-4 resonance changes with the substituent. Investigation
of the relationships of these shieldings to the calculated charge
densities in the 4,4'-disubstituted biphenyls shows that the best
overall correlations of δ_C for all positions are obtained with the
total charge. A plot of the total charge __vs.__ δ_C is shown in Figure 4. The fit is very good for C-1 and C-3,5. Points for C-2,6
are closely clustered since the variation in $\delta_{C-2,6}$ is small.
Three of the C-4 shieldings also fit fairly well, however the
points for carbons with directly bonded F, NO_2, and NH_2 fall well
off the graph. These points are in qualitative agreement with the
others, but the scatter is large, as it is for substituted carbons
in the correlation between total charge densities and ^{13}C shieldings
in substituted benzenes reported by Nelson, __et al.__[29]

Figure 5 illustrates the equivalence of substituent effects in
the 4,4'-disubstituted biphenyls and those in monosubstituted benzenes. Here the difference between the shieldings at a given carbon
in the substituted biphenyl minus the shielding of that carbon in
biphenyl is plotted against the shielding at an equivalent position
in benzene with the same substituent minus the shielding of benzene
itself. For example, δ_{C-1} (4,4'-dimethylbiphenyl) − δ_{C-1} (biphenyl)
is compared to δ_{C-4} (toluene) − δ_C (benzene). It is easily seen
that the substituent effects in the two systems are remarkably
alike.

When an intercomparison is made among the chlorinated biphenyls, it is apparent that the effects of varying chlorine
substitution on the ^{13}C shieldings are fairly small, as can be
seen from the data in Table 5. The δ_C values for carbons with
attached chlorines are consistently larger than those for protonated
carbons, reflecting deshielding by the chlorines. Carbons which
are __ortho__ to two chlorine atoms are also deshielded, although deshielding by one __ortho__ chlorine is noticeable for only some of the
PCB isomers. Protonated carbons __para__ to a chlorine substituent are
generally more shielded than those at other positions, which agrees
with benzene data. The total range of shieldings is only 16 ppm.
If the C-1 resonance, which is at lowest field for all the PCBs
studied, is excluded, the range is even narrower, 10.1 ppm.

Table 4

^{13}C Shieldings of 4,4'-disubstituted Biphenyls[a]

	δ, ppm			
Substituent	C-1	C-2,6	C-3,5	C-4
H	141.0	126.8	128.4	126.9
Cl	138.3	128.1	128.9	133.7
CH_3	138.2	126.7	129.2	136.5
NO_2	144.8	128.2	124.2	140.2
CH_3O	133.5	127.6	114.2	158.6
F[b]	136.3	128.4	115.5	162.2

[a]Approximately 0.3$\underline{M}$ in chloroform-$\underline{d}$ at 40°C.

[b]$^1J_{CF} = 246.7$, $^2J_{CF} = 21.5$, $^3J_{CF} = 7.9$, $^4J_{CF} = 2.7$ Hz.

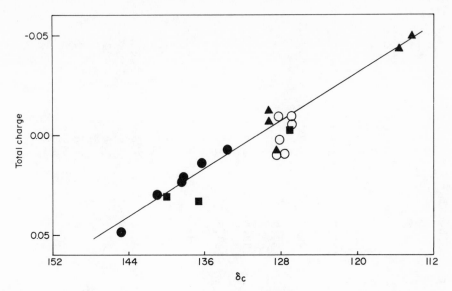

Fig. 4. Total charge vs. δ_C for 4,4'-disubstituted biphenyls.
● C-1; ○ C-2,6; ▲ C-3,5; ■ C-4.

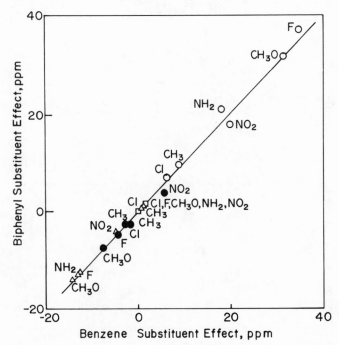

Fig. 5. Substituent effects on ¹³C shieldings of 4,4'-disubstituted biphenyls vs. those for benzenes. ● C-1; ○ C-4; △ C-3,5; □ C-2,6.

Table 5

^{13}C Shieldings of Symmetric Chlorinated Biphenyls[a]

δ, ppm

Substitution	C-1	C-2	C-3	C-4	C-5	C-6
nil	140.6	126.8	128.4	126.9	128.4	126.8
2,2'	137.6	132.8	130.7	128.7	126.0	128.9
3,3'	140.7	127.4	134.2	126.8	129.6	124.8
4,4'	137.5	127.7	128.5	132.9	128.5	127.7
2,4,2',4'	135.2	134.2	129.0	133.8	126.6	131.5
2,6,2',6'	134.4	134.3	127.4	129.8	127.4	134.3
2,4,5,2',4',5'	135.3	130.7	131.7	133.9	131.9	130.6
2,4,6,2',4',6'	134.9	135.0	127.8	132.2	127.8	135.0
3,4,3',4'	137.8	128.3	132.5	131.8	130.4	125.7
3,5,3',5'	140.4	125.1	134.9	127.9	134.9	125.1
3,4,5,3',4',5'	136.6	126.4	134.2	133.6	134.2	126.4

[a] Approximately 0.6M in 1,2-dibromoethane-d_4 at 40°C.

Nevertheless, for all the PCBs, resonances corresponding to each
set of equivalent carbons could be clearly distinguished in the
¹³C spectra. The spectra of the individual isomers are sufficiently
different from one another to permit their use as a means of identification of and discrimination between isomeric PCBs.

Since ¹³C shielding data were available for a number of chlorinated benzenes,[39] the chlorine substituent effects in these compounds were used to predict the ¹³C shieldings in the PCBs using
additivity.[36,37] A comparison of the actual chlorine substituent
effects in the biphenyls with those predictions is shown in Figure
6. For non-hindered PCBs, the points fall very close to the line
for a perfect correlation with slope +1. The scatter in these is
essentially the same as it is for the 4,4'-disubstituted biphenyl
substituent effects discussed previously (Figure 5). The presence
of chlorine substituents in the second ring, with the exception of
substituents ortho to C-1, does not significantly affect the ¹³C
shieldings in the first ring. Otherwise, greater deviations from
the 1:1 correlation would be seen, since a perfectly analogous
comparison to the chlorinated benzenes would be a comparison
utilizing δ_C values for PCBs substituted on only one ring and biphenyl itself. The deviation of the C-1 shielding in the PCBs with
ortho substituents probably results from steric hindrance to rotation about the inter-ring bond.

The ¹³C shieldings for the PCBs are grossly correlated with
the total charge densities, as is shown in Figure 7 for all the
PCBs studied which can be assumed to be freely rotating. The points
in the figure can be divided into three clusters. That on the left
side of the figure contains only points corresponding to C-1, that
at the bottom contains points for carbons with directly bonded
chlorines and the group on the right of the figure represents the
protonated carbons, with those that are ortho to a chlorine substituent being more deshielded. Thus the correlation may be considered
to be between type of substitution and total charge density, and
does not necessarily serve to distinguish between the PCB isomers.
However, for carbons without attached chlorines, the total charge
does predict the relative shieldings for an individual chlorinated
biphenyl.

The shifts in total charge at a given carbon in 4,4'-disubstit-
uted biphenyls and freely rotating chlorinated biphenyls relative to
that carbon in biphenyl are compared to the shifts in total charge
at a given carbon in the substituted benzenes relative to that in
benzene in Figure 8. As could be anticipated from the experimentally
determined substituent effects, these total charge shifts resulting
from substitution are nearly identical for the biphenyls and the
benzenes. Since the substituent effects in non-hindered biphenyls
and in benzenes have been shown to be essentially the same, these

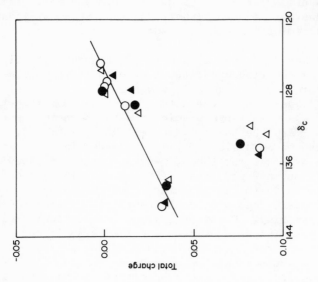

Fig. 7. Total charge density vs. ^{13}C shielding for non-hindered chlorinated biphenyls. ● 4,4'; ○ 3,3'; ▲ 3,5,3',5'; △ 3,4,3',4'. ▲

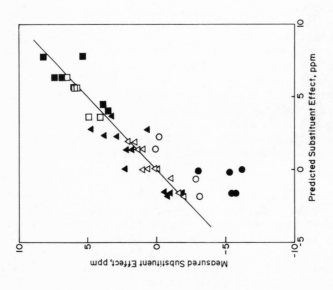

Fig. 6. Measured substituent effect vs. predicted substituent effect on ^{13}C shieldings in chlorinated biphenyls. Shaded symbols are for chlorinated biphenyls substituted ortho to C-1 or C-1'. ○ C-1; ☐ C-Cl; ○ C-1'; △ C-H.

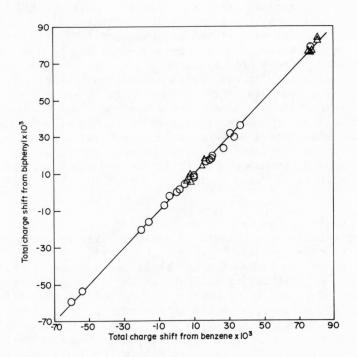

Fig. 8. Total charge shift from biphenyl for 4,4'-disubstituted biphenyls and freely rotating chlorinated biphenyls <u>vs</u>. total charge shift from benzene for analogous substituted benzenes. ○ 4,4'; △ PCBs.

two classes of compounds might be expected to have the same charge density dependences of δ_C for a series of chlorine substituents. The substituent effect on the ^{13}C shielding was compared to the calculated substituent effect on the total charge density at a given carbon for a number of chlorinated benzenes. Surprisingly, there was no correlation between the substituent effects on total charge density and on δ_C. The relationship of the ^{13}C shieldings to the π charge densities had already been investigated, and these had been found to correlate less well than the δ_C values and the total charge. Therefore, the possible interrelation of ^{13}C substituent effects in the chlorinated benzenes and PCBs with substituent effects on the σ charge densities was explored. A plot of these substituent effects for chlorinated benzenes is shown in Figure 9. The data are for all non-chlorine-substituted carbons in mono- and di-chlorobenzenes,[39] the parameter $\Delta\delta$ is the ^{13}C shielding at a given carbon in a chlorinated benzene minus the ^{13}C shielding of benzene itself and the sigma charge shift is the sigma charge density at the chlorinated benzene carbon minus the sigma charge density on a carbon in benzene. It can be seen that there is a linear relationship between the shift in the ^{13}C shielding and the shift in the σ charge density at that carbon in the chlorinated benzenes.

   Similar correlations were then sought for the chlorinated biphenyls. The relationship between $\Delta\delta$ and the shift in total charge density for the non-hindered chlorinated biphenyls was examined and found to be random, as it was for the chlorinated benzenes. It appears that no correlation exists between these two parameters in these systems. There is a correlation between $\Delta\delta$ and the shift in σ charge density, however, as shown in Figure 10. This linear dependence of $\Delta\delta$ on the sigma charge shift holds true for all carbons in the freely rotating PCBs except the carbons at positions 1 and 1'; these correspond to the points lying far below the line. Even these points appear to represent a roughly linear correlation of $\Delta\delta$ with σ charge. Their deviation from the line for the other carbons may reflect an increased π density at C-1 due to the presence of the second ring and the effects of its substituents. Interestingly, the slope of the line drawn is almost the same as the slope of the linear $\Delta\delta$ vs. sigma charge shift correlation for the chlorinated benzenes.

   In conclusion, then, one may say that the ^{1}H shieldings in the 4,4'-disubstituted biphenyls are determined by the distribution of the π charge in the aromatic rings and by ring current effects. The differences between ^{1}H shieldings in the various PCB isomers are also functions of these quantities. For each PCB isomer examined, protons ortho to a chlorine are shifted downfield relative to the other protons in the molecule. The differences in shielding of PCB protons are smaller than those for 4,4'-disubstituted biphenyls with

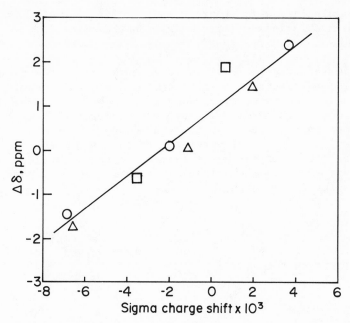

Fig. 9.   Shift of ¹³C shielding from benzene <u>vs</u>. σ charge shift from benzene for chlorinated benzenes.   △ monochloro;   ○ <u>meta</u>-dichloro;   □ <u>ortho</u>-dichloro.

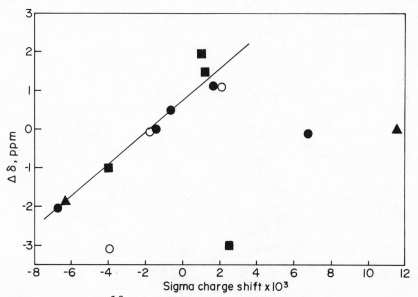

Fig. 10.   Shift of ¹³C shielding from biphenyl <u>vs</u>. σ charge shift from biphenyl for freely rotating chlorinated biphenyls.   ● 3,3';   ▲ 3,5,3',5';   ■ 3,4,3',4';   ○ 4,4'.

other than chlorine substituents. This is the result of the small
effect that a chlorine substituent has on the aromatic π electron
density.

The ^{13}C shieldings of 4,4'-disubstituted biphenyls correlate
well with the total charge densities at the carbons. For the
freely rotating PCB isomers, a general trend between total charge
and ^{13}C shielding is observed. For each isomer, there is a remarkably good correlation of total charge and ^{13}C shielding. As
in benzenes, the effects of varying chlorine substitution in biphenyls are best explained by variations in the σ charge distribution. Except for ortho-substituted biphenyls, benzene substituent
effects predict those in the biphenyls.

Acknowledgments: The assistance of Prof. J. B. Stothers in
obtaining some of the ^{13}C nmr data is greatly appreciated.
Dr. E. O. Oswald and Dr. L. A. Levy kindly provided samples of
several chlorinated biphenyls.

References

1. D. B. Peakall and J. L. Lincer, BioScience, **20**, 958 (1970);
    R. Edwards, Chem. Ind., 1340 (1971).
2. L. Fishbein, J. Chromatogr., **68**, 345 (1972).
3. See, for example, papers presented at the December 1971 Con-
    ference on PCBs, Rougemont, North Carolina, sponsored by the
    National Institute of Environmental Health Sciences, reported
    in Environmental Health Perspectives, Exp. Issue No. 1, 1972.
    (DHEW Publication No. NIH 72-218.)
4. D. Welti and D. Sissons, Org. Mag. Reson., **4**, 309 (1972).
5. K. D. Bartle, J. Ass. Offic. Anal. Chem., **55**, 1101 (1972).
6. A. R. Tarpley and J. H. Goldstein, J. Phys. Chem., **75**, 421
    (1971).
7. W. B. Smith, A. M. Ihrig and J. L. Roark, J. Phys. Chem., **74**,
    812 (1970); Y. Nomura and Y. Takeuchi, Tetrahedron Lett., 5585
    (1968); M. J. S. Dewar and A. P. Marchand, J. Amer. Chem. Soc.,
    88, 3318 (1966); D. M. Grant, R. C. Hirst and H. S. Gutowsky,
    J. Chem. Phys., **38**, 470 (1963); S. Brownstein, J. Amer. Chem.
    Soc., **80**, 2300 (1958).
8. H. Hasegawa, M. Imanari and K. Ishizu, Bull. Chem. Soc. Jap.,
    45, 1153 (1972).
9. R. Freeman, H. D. W. Hill and R. Kaptein, J. Mag. Reson., **7**,
    327 (1972).
10. G. C. Levy, J. C. S. Chem. Commun., 47 (1972).
11. G. E. Maciel and J. J. Natterstad, J. Chem. Phys., **42**, 2427
    (1965); T. D. Alger, D. M. Grant and E. G. Paul, J. Amer. Chem.
    Soc., **88**, 5397 (1966).
12. G. C. Levy, J. C. S. Chem. Commun., 352 (1972); G. C. Levy,

        J. D. Cargioli and F. A. L. Anet, J. Amer. Chem. Soc., **95**,
        1527 (1973).

13.   O. Gropen and H. M. Seip, Chem. Phys. Lett., **11**, 445 (1971).

14.   M. J. S. Dewar and A. J. Harget, Proc. Roy. Soc. Lond. A, **315**,
        443 (1970); A. Golebrewski and A. Parczewski, Z. Naturforsch.,
        A, **25**, 1710 (1970); B. Tinland, Theor. Chim. Acta, **11**, 452
        (1968); G. Casalone, C. Mariani, A. Mugnoli and M. Simonetta,
        Mol. Phys., **15**, 339 (1968); N. C. Baird and M. J. S. Dewar,
        J. Amer. Chem. Soc., **89**, 3966 (1967); I. Fischer-Hjalmars,
        Tetrahedron, **19**, 1805 (1963); H. Suzuki, Bull. Chem. Soc. Jap.,
        32, 1340 (1959); F. Adrian, J. Chem. Phys., **28**, 608 (1958).

15.   H. Suzuki, Bull. Chem. Soc. Jap., **32**, 1350 (1959); **32**, 1357
        (1959).

16.   H. Suzuki, Bull. Chem. Soc. Jap., **33**, 109 (1960).

17.   M. Karplus and J. A. Pople, J. Chem. Phys., **38**, 2803 (1963).

18.   E. Farbrot and P. N. Skancke, Acta Chem. Scand. **24**, 3645
        (1970).

19.   J. R. Pedersen, Acta Chem. Scand., **26**, 3181 (1972).

20.   K. E. Howlett, J. Chem. Soc., Pt. 1, 1055 (1960).

21.   D. P. Santry and G. A. Segal, J. Chem. Phys., **47**, 158 (1967);
        D. P. Santry, J. Amer. Chem. Soc., **90**, 3309 (1968).

22.   J. J. Kaufman and R. Predney, Int. J. Quantum Chem., 231
        (1972).

23.   J. R. Sabin, D. P. Santry and K. Weiss, J. Amer. Chem. Soc.,
        94, 6651 (1972).

24.   A. Rauk, J. D. Andose, W. G. Frick, R. Tang and K. Mislow,
        J. Amer. Chem. Soc., **93**, 6507 (1971).

25.   N. K. Wilson, M. Anderson and J. B. Stothers, unpublished
        results.

26.   J. A. Pople and D. L. Beveridge, Approximate Molecular Orbital
        Theory, McGraw-Hill, New York, 1970; J. A. Pople, Accounts
        Chem. Res., **3**, 217 (1970).

27.   R. J. Pugmire and D. M. Grant, J. Amer. Chem. Soc., **93**, 1880
        (1971).

28.   P. Lazzeretti and F. Taddei, Org. Magn. Resonance, **3**, 282 (1971).

29.   G. L. Nelson, G. C. Levy and J. D. Cargioli, J. Amer. Chem.
        Soc., **94**, 3089 (1972).

30.   (a) Program 111, Quantum Chemistry Program Exchange, Indiana
            University.
        (b) Program 141, Quantum Chemistry Program Exchange, Indiana
            University.

31.   L. M. Jackman and S. Sternhell, Applications of Nuclear
        Magnetic Resonance Spectroscopy in Organic Chemistry, Pergamon,
        New York, 1969.

32.   H. P. Figeys and R. Flammang, Mol. Phys., **12**, 581 (1967).

33.   R. E. Mayo and J. H. Goldstein, Mol. Phys., **10**, 30 (1966);
        G. E. Johnson and F. A. Bovey, J. Chem. Phys., **29**, 1012 (1958).

34.   Y. Nomura and Y. Takeuchi, Tetrahedron Lett., 5665 (1968).

35.   J. F. Sebastian and J. R. Grunwald, Can. J. Chem., **49**, 1779

(1971); J. M. Haigh and D. A. Thornton, Tetrahedron Lett.,
2043 (1970); P. Lazzeretti and F. Taddei, Tetrahedron Lett.,
805 (1970); T. K. Wu and B. P. Dailey, J. Chem. Phys., 41,
2796 (1964); J. C. Shug and J. C. Deck, J. Chem. Phys., 37,
2618 (1962); G. Fraenkel, R. E. Carter, A. McLachlan and
J. H. Richards, J. Amer. Chem. Soc., 82, 5846 (1960).

36. J. B. Stothers, Carbon-13 NMR Spectroscopy, Academic Press,
     New York, 1972.

37. N. K. Wilson and J. B. Stothers, "Stereochemical Aspects of
     ^{13}C NMR Spectroscopy", in Topics in Stereochemistry, Vol. 8,
     E. Eliel and N. L. Allinger, ed., Wiley-Interscience, New York,
     1973. (In press).

38. G. C. Levy and G. L. Nelson, Carbon-13 Nuclear Magnetic
     Resonance for Organic Chemists, Wiley-Interscience, New York,
     1972.

39. A. R. Tarpley and J. H. Goldstein, J. Phys. Chem., 76, 515
     (1972); A. R. Tarpley and J. H. Goldstein, J. Mol. Spectrosc.,
     39, 275 (1971).

PROTON FOURIER TRANSFORM NMR AS A TRACE ANALYSIS TOOL:

OPTIMIZATION OF SIGNAL-TO-NOISE

F.H.A. Rummens

Department of Chemistry, University of Saskatchewan

Regina Campus, Regina (Sask.) Canada

ABSTRACT

An experimental study has been made of all the factors which influence the Signal-to-Noise ratio in FT NMR. In particular the achievable Signal-to-Noise per time unit is discussed. A detection limit of 1 ppm (1 μg) or a reasonable spectrum at 10 ppm (10 μg) is achievable in 1 hour of experimentation time. A further improvement by a factor of 10 in detection limit appears feasible, at the sacrifice of very long experimentation time but the major obstacle is provided by proton-carrying solvent impurities. In practical applications the same problem gives rise to the need for highly specific extraction and clean-up procedures. A general description of the Puls Fourier Transform technique is given and a general method for optimizing towards maximum Signal-to-Noise per-time-unit is evolved.

CW VERSUS FT NMR SPECTROSCOPY

A traditional NMR experiment proceeds as follows; the strong and constant magnetic field H_0 attempts to orient the nuclear dipoles along the direction of H_0 (z-axis). For nuclei of spin $\frac{1}{2}$, such as protons, there are *two* possibilities for the z-component of the magnetization; either with or against the direction of the field. There will be more nuclei "lined up" with H_0 since the latter configuration has the lower energy. It is the excess magnetization in the positive z-direction that eventually makes detection of NMR possible. The uncertainty principle forbids the exact line-up of the nuclear dipoles; a certain angle between the dipole vector and H_0 must be maintained. As a result, the nuclei (or rather their dipole-vectors) start a precession motion around the z-axis with a precession frequency $\nu_{pr} = \gamma H_0$ (see fig. 1) where γ is the gyromagnetic constant of the nucleus in question. In the next step radio frequency electromagnetic radiation is introduced into the sample by means of a cylindrical coil, wrapped around the sample-tube; this coil is perpendicular to H_0 (y-axis). This r.f. radiation contains as a component a magnetic field of strength H_1 whose direction rotates in the xy plane with frequency ν_1 and with the same rotational sense as the precession of the nuclei. The frequency ν_1 can be changed (frequency scan). If and when ν_1 becomes equal to ν_{pr} the field H_1 applies a lasting torque on the nuclear dipoles, caus-ing them to flip over to the higher energy state. The energy required for this is absorbed from the electromagnetic field. This can be detected electronically, thereby producing the resonance peaks in the spectrum. In this technique the field H_1 (of frequency ν_1) is constantly "on"; hence the name "continu-ous wave" or CW NMR. The major disadvantage is that while ν_1 is scanned over the entire sweep width or spectral range Δ (about 1000 Hz for protons) it is really only effective during the time it takes to go through a line of line width $\Delta\nu_{\frac{1}{2}}$. The wastage of time is therefore a factor of about $\Delta/\Delta\nu_{\frac{1}{2}}$ or about 2000 if $\Delta\nu_{\frac{1}{2}} \approx 0.5$ Hz.

Puls Fourier Transform NMR aims to recoup this loss of time. Fourier Transform NMR was introduced to enable measurement on nuclei of low inherent sensitivity and (or) low isotopic abundance

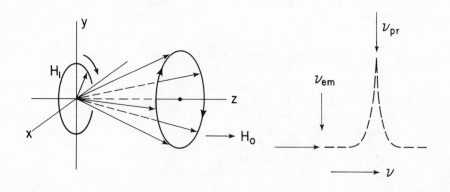

$$\nu_{precession} = \gamma H_0$$

Fig. 1 Principle of Continuous Wave (CW) NMR.

(such as ^{13}C). The same kind of sensitivity gain can also be
obtained, however, for sensitive and abundant nuclei such as protons; the gain is then commonly utilized to either reduce the
required time or to lower the detection limit. Quite a few simple explanations of the Fourier Transform method [1,2] as well as
more advanced treatments [3,4] are available in the literature.

In this paper all the factors that influence Signal-to-Noise
have been experimentally investigated and tested against theoretical predictions. In the Puls FT technique the strength of H_1
is first boosted by a power amplifier and is then via a puls
unit loosed on the sample for a very short time (typical pulswidth is 10 μsec.). If r.f. power and pulswidth are set properly
this results in the magnetization being turned 90° from the z
direction into the xy plane. A detector coil (around x-axis
in a cross-coil probe) then would see a strong signal of all the
nuclei together since the r.f. power is large enough to irradiate
all the nuclei irrespective of their chemical shift. (In fact
the frequency of the H_1 field is deliberately chosen to be just
outside the spectral range either at the high field or the low
field end). This signal will subsequently decay in time; the
composite transverse relaxation time T_2 will cause the xy magnetization to decay to zero, while at the same time the longitudinal relaxation time T_1 will build up the z magnetization until
the latter reaches its thermal equilibrium value again. When
this has been achieved a new puls may be produced to start the
procedure again. The observed decaying signal is not a simple
declining function; the detector circuity contains a reference
which is synchronized with ν_1. The signal frequency is somewhat
different however, so that its (decaying) signal will go in and
out of phase with the reference frequency ν_1 of the detector.
This causes an oscillating signal pattern (see fig. 2) which is
commonly referred to as the free induction decay (FID) signal. Of
course, the frequency difference mentioned above is precisely the
distance in the frequency spectrum between the frequency ν_1 and
the frequency of the sample signal. As the FID decays in time one
could pick out the period of this frequency difference by measuring the time difference between every second traversal through
the baseline. Inversion of that period would then give the frequency difference. Not only would this be a tedious job, but it
would become almost impossible if the spectrum contained more than
one line, so that the FID would be a very complex function of
superimposed individual FID's. However, there is a mathematical
technique, called Fourier Transformation, which, when applied to
the FID "time domain" function, produces the desired "frequency
domain" spectrum.

Figure 3 shows the time sequency of events. After the puls
there is a short delay time to give the electronic circuitry

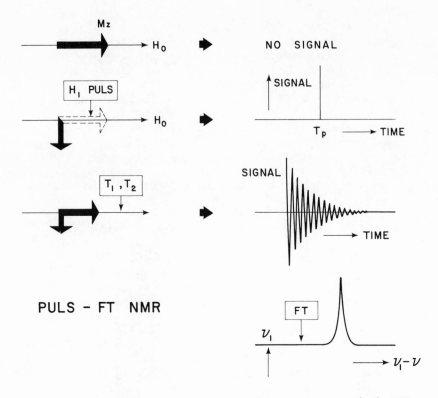

Fig. 2   Principles of Puls-Fourier-Transform (FT) NMR.

the time to recuperate from the effects of the heavy H_1 puls.
The actual data collection may take several seconds. An
analog-to-digital converter (ADC) translates the (analog)
voltage into digits which are then stored into a computer
memory. If, say, this memory is 4K large (i.e. there are
$2^{12} = 4096$ "addresses" or places to store numbers in) and data
collected over 2 sec then the ADC spends 2/4096 or about 488 μsec
in each "address". This time per address is called the dwell
time; its reciprocal is the sampling rate (2048 addresses per
second in the above example). This sampling frequency is also
important because it determines the frequency range of the
spectrum. Since each sine wave needs at least two points per
period to be digitized without ambiguity, it follows that no
unambiguous spectral frequency larger than half the sampling frequency can be measured. Therefore, in the above example the spectral range was ½ x 2048 = 1024 Hz.
    When the computer fourier-transforms 4K of FID data it pro-
duces two spectra each of 2K points, the so-called "real" and

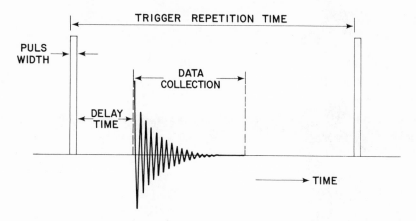

Fig. 3   Schematic Time Sequence of Puls Experiment

"imaginary" spectrum (see fig. 4). These two spectra differ
only in phase information but are otherwise identical. By making
a phase-correction (also carried out by the computer) it is
possible to convert the "real" part to a true absorption or
v-mode spectrum. At the same time this makes a pure dispersion
or u-mode spectrum out of the "imaginary" part. Usually only
the absorption spectrum is converted back into analog form and
displayed on a recorder. The bottom part of fig. 4 shows the
so-called magnitude spectrum defined as the square root of
$u^2 + v^2$. This has the effect of increasing the signal-to-noise
ratio by a factor $\sqrt{2}$.

INSTRUMENTAL

A Bruker HX-90 spectrometer was used throughout. The instrument was equipped with a B-SV3P puls unit, a B-LV80 90 MHz power
amplifier and a Nicolet 1083 computer with 12K of memory. As a
test sample 2,4 dichlorophenoxyacetic acid at 0.5% in $CDCl_3$
(99.8% isotopic purity) was chosen. In particular the Signal/
Noise ratio of the $-OCH_2-$ protons were studied. This signal
appears to have a natural line width of about 0.6 Hz. The choice
of a not-too-narrow line was deliberate, because it was felt that
such a linewidth is more representative of actual circumstances
(whether caused by long range coupling, dissolved oxygen or non-
optimum magnet homogeneity) than any selected very narrow line.
To the sample about 3% v/v C_6F_6 was added; the ^{19}F signal of this

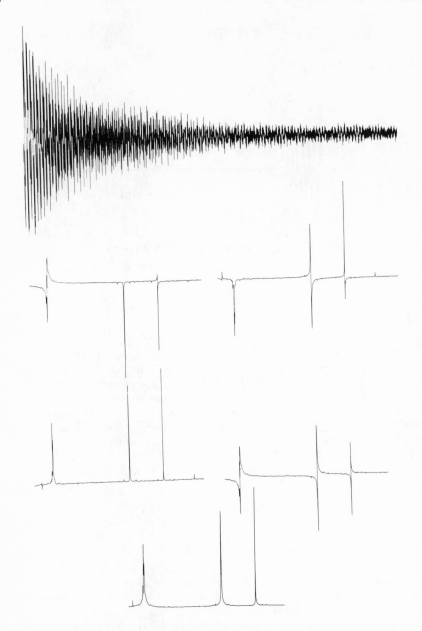

Fig. 4 Top; Free induction signal of Tolueme (plus TMS) collect-
        ed in 4k of memory. Total collection time 2 sec. Second
        row; "real" and "imaginary" part of frequency spectrum,
        immediately after Fourier transformation. Third row;
        same as above after phase correction. Bottom; the
        "power" or "magnitude" spectrum.

was used to provide a lock-signal to maintain long term stability.
Also about 1% TMS was added to provide an easily observable FID
signal, that could be monitored on the oscilloscope for each individual puls. The Signal/Noise ratio is defined as follows;
(see also figure 10).

$$S/N = 2.5 \times \frac{\text{peak height measured from noise centre}}{\text{peak-to-peak noise level}}$$

Much can be said about the factor 2.5 in the above definition [3]
let it be sufficient to say that it constitutes a conservative
theoretical limit and that in practice S/N = 1 corresponds to a
signal that can indeed just be observed by visual inspection of the
spectrum. By convention one also has then defined the detection
limit as the situation where S/N is unity.

SIGNAL-TO-NOISE OPTIMIZATION

        In this section a discussion will be given of how the various
instrumental parameters can be optimized to produce the maximum
Signal/Noise ratio. In theory one would expect that an infinitely
large S/N is obtainable by pulsing an infinite number of times and
stacking all these FID's together. Clearly there are limits here,
imposed by long term instabilities of the instrument, the cost of
tying up the instrument for one experiment and, not the least, the
impatience of the experimenter who wants results fast and plenti-
fold. In spite of these limitations it is nevertheless true that
"time" is an important part of any discussion on S/N. It would be
more appropriate to talk about S/N *per time unit* as we shall indeed do in this paper. In general, one can expect everything
else being equal, that S/N will increase with the square root of the
total experimentation time. For example, where we said above that
FT would be faster for the same S/N then CW by a factor of $\Delta/\Delta\nu_{\frac{1}{2}}$,
or about 1000/0.6 = 1600, in terms of S/N we then would expect FT
to be better by a factor of 40 on the basis of equal total time.
In actual fact an improvement of somewhat less than that is to be
anticipated, because CW NMR can be done considerably faster than
the steady-state scanning conditions would require, without too
much loss of signal-to-noise. It is perhaps realistic to expect
a S/N improvement of no more than 10x on the basis of equal time.
Translated back in time this would be equivalent to a 100 fold
reduction in time; this in turn could be the difference between an
acceptable 1 hr 40 min per FT or a totally unacceptable 7 days for
CW.

R.F. Power and Puls Width

To a large extent it is the product of r.f. power (H_1 strength) and pulswidth that is important. It is this product that determines the degree by which the magnetization is turned away from the z-axis and towards the xy plane. At a given pulswidth too little H_1 will not give enough xy magnetization, too much H_1 may result in a 180° puls (i.e. no xy magnetization) or even in total destruction of any magnetization (saturation). Figure 5 shows that for 12.5 µsec of pulswidth there is a rather sharply defined optimum for 45 decibels of r.f. attenuation. The equivalence of H_1 power and pulswidth is (at least in part) demonstrated by varying pulswidth at 45 dB of H_1 power. Not surprisingly the optimum falls near 12.5 µsec. Within rather wide limits the choice of pulswidth is not important; it has to be considerably larger than the "rise-time" of the puls (about 0.5 µsec in the present case) to guarantee a squared-off puls. On the other side pulswidth has to be considerably smaller than the smallest relaxation time so that no appreciable relaxing-back takes place during the puls. There is only one limitation on the strength of H_1 namely that it be strong enough so that all nuclei see *effectively* the same field strength resulting in one and the same flip angle for all nuclei, irrespect-

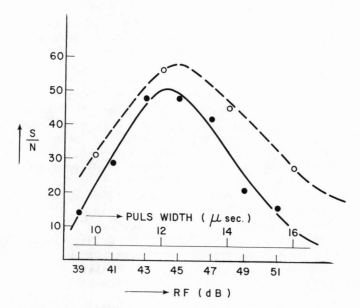

Fig. 5   Solid line S N as function of Db of r.f.
attennuation at 12.5 µsec of puls width. Dotted line;
S/N as function of puls width t_p at 45 Db of attennuation.

ive of the $\nu_{pr} - \nu_l$ differences. The relation between magnetization flip angle θ, puls width t_p and r.f. field strength H_1 is
given by:

$$\theta = t_p \nleftarrow H_1 \qquad\qquad (2)$$

It may be noted that although $\theta = \pi/2$ gives the strongest initial
FID signal, the return from $\theta = \pi/2$ to $\theta = 0$ (and to additionally
restore the z-magnetization) takes a long time (about $4T_1$ sec).
For maximum S/N per time unit it may be advantageous to use a θ
considerably smaller than $\theta/2$ (About 20-30°). The loss in signal
per puls is more than made up by the larger number of pulses. If
indeed such shorter trigger times are used, either H_1 or t_p may
have to be re-optimized.

Number of Pulses

Because of the virtually absolute synchronism maintained by
the computer clock between the various time-based events it is
possible to produce many FID signals and stack these on top of
each other in the computer memory. In this way signal strength
will go up proportionally with N (number of pulses). The noise
will also go up but because of the random nature of this noise a
certain averaging out takes place. For truly random noise the
noise will increase proportionally with $\sqrt{N}$. Therefore, S/N
should increase with $N/\sqrt{N} = \sqrt{N}$. Since total experiment-time t
goes up linearly with N, one might also say that S/N is proportional to $\sqrt{t}$.

Figure 6 shows that S/N increases with $\sqrt{N}$ even over long
times. In this experiment the trigger time was 2 sec and the largest N was 4096 for a total time of 2 hrs 16 min. Although not all
parameters were optimized at this stage, a S/N = 312 was achieved,
which corresponds to a detection limit of 16 ppm (μg/ml). In the
above experiment the total data collection time was just under 2
sec (i.e. just below the trigger time). In a separate experiment
the trigger time was lengthened, while keeping collection time
constant. This had the advantage that the nuclei could more
completely relax back to their thermal equilibrium. However, it
was found that for a constant N this barely increased S/N, or put
differently, that S/N per time unit went down almost proportionally with the square root of trigger time.

Trigger Time and Collection Time

From the above the suggestion follows that perhaps it is advantageous to puls fast even though it may mean incomplete restora-
tion of thermal equilibrium before the next puls arises. It is

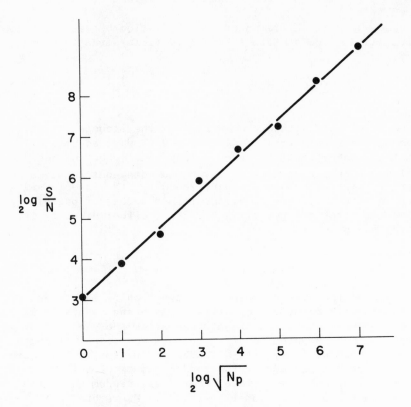

Fig. 6 Signal-to-Noise as function of the
         total number of pulses.

mandatory, however, that the data collection of one FID is completed before the next puls. A new complication arises here. If
it is desirable to keep the spectral range constant (e.g. 1050 Hz)
this forces the dwell time to be constant (476 μsec) as well
(See above). If now the total collection time is to be reduced,
the only possible solution is to reduce the memory size in which
the FID data are to be stored. For example, 8K of core
(i.e. 8192 addresses) at a dwell time of 476 μsec means a total
collection time of 3.9 sec allowing a 4.0 sec trigger time. On
the other hand a 1K data core requires only 0.49 sec, allowing
trigger times as small as 0.5 sec. The disadvantage of small core
is of course extra line broadening; 1K of addresses to cover 1050
Hz means 1 Hz of digitizing width which adds 2 Hz to the natural
line width after the FT is completed, reducing the line height
in proportion. For very narrow lines this digitizing broadening
soon becomes very serious, disallowing short trigger times. This
is one way to show that increased line width or its corresponding
loss in signal strength can be recovered with suitable FT techni-

ques to a large extent. It is analogous to the gain in S/N per time unit that can be achieved in CW spectroscopy by scanning faster than steady conditions dictate (whereby also line-broadening is incurred).

TABLE I

S/N FOR 16 PULSES
DWELL TIME 476 μ SEC, SW = 1050 Hz

COLLECTION TIME (SEC)	CORE SIZE (K)	TRIGGER TIME (SEC)			
		4	2	1	0.5
0.5	1	50	40	32	25
1	2	43	33	27	
2	4	32	21		
4	8	19			

Table 1 gives some results for 16 pulses at various trigger times and collection times. At constant collection time the larger trigger time has the larger S/N. This reflects the more complete relaxation. At a given trigger time the S/N goes down with increased collection time; as the FID decays one ends up collecting more noise than signal. At very short collection times a reduction in S/N would be expected because the digitization broadening would become excessive.

In Table 2 the data are recalculated towards an equal total time basis (assuming S/N proportional to $\sqrt{t}$). This is the more relevant basis; it produces the perhaps unexpected result that the best S/N per time unit is obtained for very short collection and trigger times, in spite of the small memory core (1K) and its consequent broadening. At these short trigger times the FID has by no means decayed completely when the next puls arrives. The magnetization flip angle θ oscillates between a minimum and a maximum value, never quite zero, never quite 90° and a pseudo-stationary state will set in. Taken to its extreme one might imagine that a much decreased trigger time with weaker pulses might create an equilibrium state with even better S/N. The consequences of such a technique are not entirely clear at this moment.

TABLE 2

S/N FOR 64 SEC
DWELL TIME 476 μ SEC,   SW = 1050 Hz

COLLECTION TIME (SEC)	CORE SIZE K	TRIGGER TIME (SEC)			
		4	2	1	0.5
0.5	1	50	57	64	70
1	2	43	46	53	
2	4	32	30		
4	8	19			

Delay Time

A delay time is introduced to enable the receiver electronics
to recover from the effects of a strong transmitter field H_1.
With small or zero delay the spectrum may show spurious spikes.
A certain amount of signal is lost in this way but as long as the
delay time is very short in relation to the total decay time,
this effect is negligeable (see figure 7). As a rule the dealy
time is set approximately equal to one dwell time (several hundred
μ sec) so that actually only one bit of information is lost.

Digital Filtering

The FID shows a strong signal at the beginning but after several seconds there may be more noise than signal. Inclusion of
the latter data may in the end in fact decrease the S/N ratio
in the frequency spectrum. Cutting off the data collection at a
point where the FID signal is still strong leads to broadening of
the lines. An alternate technique is to multiply the FID with a
weighting factor of the form $\exp(-i\tau/n)$ where n is the total
number of data points in the digitized FID, τ is a time constant
and i the index number of each data point starting with zero.
Therefore, the first data point is multiplied by $\exp(o) = 1$,
subsequent data points by gradually decreasing values, and the
last data point by $\exp(-\tau)$. The calculation is carried out on
the computer and the weighted FID so obtained is then fourier-
transformed. Figure 8 shows that for an 8K FID collected over

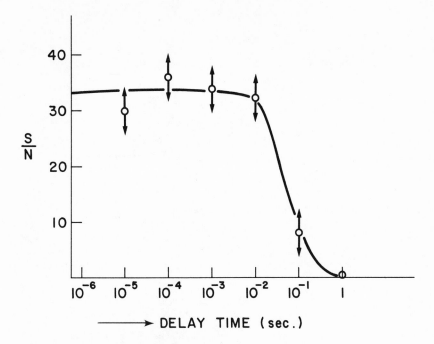

Fig. 7 Signal-to-Noise as function of Detector Delay
Time

4 sec a considerable S/N improvement may be obtained. The results indicate an optimum at around τ= 3 sec. Longer τ's result in more broadening than can be made up for by stronger filtering. It may be argued that this kind of S/N improvement is largely illusory. Indeed, as was discussed above a 4 sec collection time is not the best in the first place. As the second curve in figure 9 shows a collection time of 2 sec over 4K starts off already on a better S/N and digital filtering adds little to that. On a basis of equal number of pulses, digital filtering and reduced collection time are equivalent, but since on a total time basis reduced collection time produces larger S/N, no advantage is then obtainable by digital filtering.

Cut-Off Filter and Spectral Width

It was already mentioned in the foregoing that given a certain rate at which the FID is sampled (i.e. converted to digital points) that this limits the spectral width (in Hz) to half the sampling frequency. This so-called Nyquist frequency arises out of the fact that

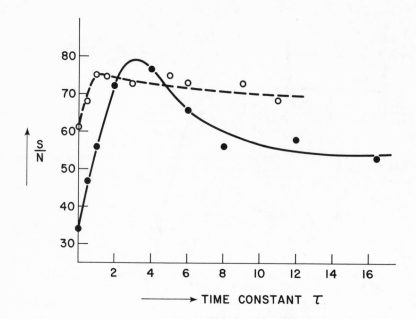

Fig. 8   Signal-to-Noise as a function of Digital
         Filtering Time Constant τ(sec).   Solid
         line; 4 sec of collection time into 8k
         of memory. Dotted line; 2 sec of coll-
         ection time into 4k of memory.

each frequency required at least two points per period for an
unambiguous digital representation. If there were another resonance line with an amount Δν beyond this limit, the digitizing then leads to a *double* solution *viz* ±Δν   from this limit.   In
other words an extra line appears an amount Δν lower than the Nyquist limit. This 'folding-back' is not only a nuisance, because of
sorting-out problems, but it is important to note that all noise
also gets folded back into the spectral range of interest, thereby
decreasing S/N. To avoid this from happening the computer has
settable sharp cut-off filters. These are set at a frequency cut-
off just beyond the spectral width. Table 3 gives some experimental results. The conditions were; dwell time 1908 μsec, spectral
width 250 Hz, memory 2K, trigger time 2 sec, 8 pulses. It is seen
that for 0.2 to 0.4 kHz filter the S/N is virtually constant.
For 0.5 or 0.6 kHz of filter frequency cut-off all the noise from
250 Hz upwards is folded back over the entire spectral range,
reducing S/N by a full factor of two (and not a factor of $\sqrt{2}$ as
one might intuitively expect). For a filter cut-off of 0.1 kHz
the signal of the resonance line itself was severely filtered down,
resulting in decreased S/N. The right hand portion of figure 10
illustrates the effect of improper filtering.

TABLE 3

S/N AS FUNCTION OF FREQUENCY CUT-OFF FILTER

FILTER (k Hz)	0.6	0.5	0.4	0.3	0.2	0.1
S/N	13	14	29	27	30	11

The spectral width of 250 Hz in the above example leads to the oft-asked question whether it is not advantageous to just measure part of the spectrum. Contrary to CW NMR where such reduction leads to a saving of time, a *loss* of time is incurred in FT NMR. Suppose one reduces the spectral width by half; this means that the FID sampling rate has to be halved or that the dwell time has to be doubled. If one maintains the same core size the total collection time then also doubles and provided the original trigger time was set just beyond the original collection time, the trigger time has to be doubled also. A larger collection time usually means increasingly more collection of noise. Because of this the S/N will go down, even for a constant number of pulses. On an equal time basis the S/N reduction is even more severe. The only off-setting factor is the decreased digitizing broadening that goes with smaller spectral width in constant core size. Table 4 gives some illustrative results. These results beg the question whether S/N could not be increased by *decreasing*

TABLE 4

S/N AS FUNCTION OF SPECTRAL WIDTH (SW)
AT 4k MEMORY

DWELL TIME (μ SEC)	SPECTRAL WIDTH (Hz)	CUT-OFF FILTER (k Hz)	COLLECTION TIME (SEC)	S/N	
				8 PULSES	128 SEC.
476	1050	1.2	2	38	107
952	525	0.6	4	28	56
1908	262	0.3	8	24	34
3816	131	0.2	16	19	19

the dwell time. If for example the dwell time were 238 μsec the
collection time for 4K would be only 1 sec. Unfortunately, this
would automatically (at least in the Nicolet 1083 computer) raise
the spectral width to 2100 Hz, the upper half of which would contain no signals; and worse even, the digitizing broadening would be
doubled to 1 Hz. The solution to this would be either to provide
double the core size (which is very costly) or to abandon the automatic coupling between dwell time and spectral width (which would
require only a minor modification in the software). There is, however, an additional disadvantage to reduced spectral width. It
usually requires the puls frequency ν_1 to be set somewhere in the
middle of the spectrum rather than at one of the ends. It is
well known that the fourier transformation cannot differentiate
between positive and negative frequency differences. In fact,
it produces two spectra mirror-imaged around the ν_1 position.
Normally, with ν_1 at one end, the mirror-image spectrum is simply
ignored but when ν_1 is in the middle of the spectrum these reflected
spectral portions may interfere very much. The left hand portion
of figure 9 illustrates this. Note that reflected or folded-
back lines usually have a completely different phase.

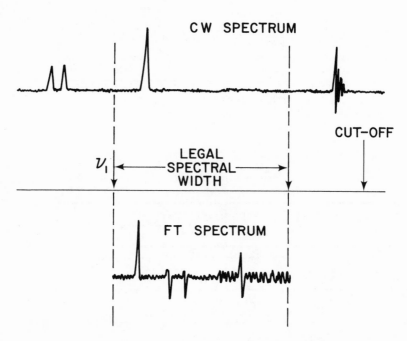

Fig. 9   Folding-back due to improper cut-off filtering
         (right-hand-side) and due to placing of frequency
         of H_1 field inside the spectrum (at left)

Digitizing Resolution

Fig. 10 shows the result on a 20 ppm 2,4D solution. At this stage it would have been possible to improve the S/N at least 10x by using 0.5 sec pulses (about a factor 2, see Table 2), by using 10 mm rather than 5 mm sample tube (about a factor of 4) and by using the "magnitude" mode (a factor of 1.4). Yet it would not have been possible to measure a more dilute solution.

The solution contained about 1000 ppm TMS. Taking into account that TMS has 12 equivalent protons versus only two for the $O-CH_2$ signal of 2,4D the signal of TMS is expected to be about 300 times stronger than that of 2,4D. In setting up the digitizer two precautions have to be taken. Firstly the vertical digitizer resolution is set to a maximum. In the present case this was $2^9 = 512$ bits; the digitizer is then prepared to subdivide the largest possible signal (voltage) that will just fit into the word length of one address into 512 parts. It will therefore store that signal with a maximum precision of $\pm 1/512$ (or $\pm 0.2\%$). Secondly, the actual number of counts per volt of signal is adjusted so that indeed the strong initial portion on the FID just about fills the word length of the digitizer. This has as a consequence that any systematic signal of a strength equal or less than 0.2% of the strongest signal will register only ± 1 count

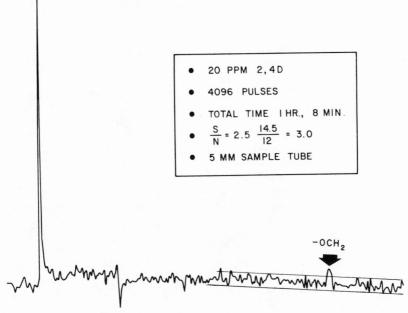

- 20 PPM 2,4D
- 4096 PULSES
- TOTAL TIME 1 HR., 8 MIN.
- $\dfrac{S}{N} = 2.5\ \dfrac{14.5}{12} = 3.0$
- 5 MM SAMPLE TUBE

$-OCH_2$

Fig. 10   Exceeding of allowable dynamic concentration range results in distortion and reduction of solute signal (see text).

after digitization; in other words it will be considered as random noise. In the present example the 2,4D signal (whose FID is
of course superimposed on that of TMS) will register only 1 or 2
counts, barely enough to be counted as systematic. It would have
been easy enough to eliminate the TMS but other signals may then
take over the role of the strongest signal. The 0.2% CHCl₃ in the
99.8% CDCl₃ causes a limit of 4 ppm as lowest possible detectable
concentration.

The above points to what is perhaps the severest drawback of
proton FT NMR. Solvent and sample should be free of proton-carrying impurities to a signal level of no more than 100 times of the
weakest solute level one want to detect even when such signals
are at non-interfering frequencies. The emphasis here is on the
weakest solute line; otherwise the stronger lines might be truly
represented while weaker ones could be totally absent which would
lead to insurmountable interpretation problems.

FINAL PARAMETRIZING AND CONCLUSIONS

It usually is of advantage to start with maximum homogeneity,
even though in the final analysis it is of advantage to choose
small memory size which leads to digitizing broadening.

A short FID collection time is recommended. Probably 0.5
sec is the optimum for proton FT since at lower collection time
the digital peak broadening becomes excessive.

Pulswidth and r.f. strength need to be optimized carefully.
An approximate optimization can be done at any condition, but
after the collection time has been chosen, this has to be repeated.
In a separate experiment it was determined that at 4 sec collection
time 45 dB provided the optimum. At 0.5 sec collection time,
however, the optimum was at 45 dB (i.e. only 25% less power). Non-
adjustment would have led to a loss of a factor of two in S/N.

A sample tube diameter of preferably 10 mm should be used.
Although this leads to some loss of homogeneity, this is almost
negligible relative to the self-imposed digital broadening.

Digital filtering is of little or no advantage in S/N per
time unit experiments.

Under the above conditions it is possible to get a decent,
fully interpretable spectrum of 10 μg (10 ppm) of solute in about
one hour of total experiment time.

Progress beyond this limit is possible at the expense of time
or by the use of special puls sequences such as SEFT and DEFT

[4].   Decoupling techniques to collapse complex signal patterns into single lines may be helpful on occasion. In the same vein the chemical conversion to derivatives with a high number of equivalent protons (such as $-Si(CH_3)_3$) may be helpful. The pragmatic limit is formed however, by the always present proton impurities in solvent or sample.

LITERATURE REFERENCES

J.W. Cooper
    "An Introduction to Fourier Transform NMR and the Nicolet 1080
    Data System."
    Nicolet Instrument Corp. Madison (Wisc.)   1972.

H, Hill, R. Freeman
    "Introduction to Fourier Transform NMR" Varian Associates
    Palo Alta (1970).

R.R. Ernst
    Advances in Magnetic Resonance (Ed. J.S. Waugh) Vol. 2
    Academic Press (1970).

D.G. Gilles, D. Shaw
    Annual Reports on NMR Spectroscopy (Ed. E.F. Mooney) Vol 5A
    pp 560-630 Academic Press 1973.

NUCLEAR MAGNETIC RESONANCE STUDIES OF THE BEHAVIOR OF DDT IN MODEL MEMBRANES

R. Haque and I. J. Tinsley

Department of Agricultural Chemistry and Environmental
Health Sciences Center, Oregon State University
Corvallis, Oregon   97331

INTRODUCTION

DDT [1,1,1-trichloro-2,2-bis(p-chlorophenyl) ethane] and
related compounds have been used extensively as insecticides.
More recently these compounds, particularly DDT and DDE [1,1-
dichloro-2,2-bis(p-chlorophenyl) ethene], have been the center of
a heated controversy in relation to their environmental behavior.
Both compounds are distributed widely in the environment, tend to
be concentrated in food chains and there is evidence that certain
wildlife populations may be affected as a consequence. Despite
the continued interest in these compounds over a number of years,
the molecular basis of their toxic action is not understood.

The effect of DDT on the central nervous system, more specifically on axonic transmission, is well documented (1-3). The
changes in action potential are associated with impaired ion (K^+,
Na^+) efflux and an effect of DDT on membrane permeability is
suggested. This suggestion would be supported by the observation
of Hilton and O'Brien (4) that DDT will block the action of
valinomycin on a lecithin-decane bilayer.

From extensive studies of DDT and related compounds Holan has
defined the stereochemical requirements for biological activity
(5,6). Other studies at the molecular level have not been particularly definitive. O'Brien and Matsumura (7) have demonstrated
that DDT will bind to components of cockroach nerve and on the
basis of spectral data the formation of a charge-transfer complex
has been suggested as a possible explanation for the binding.

Although the interpretation of this spectral data has been
questioned (8) other studies have also indicated that DDT may
form charge-transfer complexes with tetracyanoethylene (8) and
aromatic compounds (9,10).

An understanding of the molecular basis for the action of
DDT will require definition of its interactions with component
molecules of the affected systems and a demonstration of the
structural and functional changes which could result from such
interactions. With DDT being a very hydrophobic molecule and
with the suggestion of membrane involvement, a study of the
interactions of DDT with complex lipids seemed warranted. We
have used nuclear magnetic resonance (n.m.r.) techniques in
exploring the structural and thermodynamic aspects of such
interactions (11-13). Initial studies were carried out in non-
aqueous systems while more recent investigations have been
conducted in aqueous dispersions of phospholipids; the latter
systems are more respresentative of the cellular environment.
Characteristics of the binding of DDT and related compounds
to complex lipids will be summarized and the relationships between
binding and biological activity will be discussed. It will also
be demonstrated that interaction studies can provide information
on phospholipid configuration.

STUDIES IN NON-AQUEOUS SYSTEMS

Proton magnetic resonance (p.m.r.) spectra have been recorded
on a Varian HA 100-MHz spectrometer at 33°C. Chemical shift values
were measured using tetramethylsilane as a lock standard. Spectro-
grade chloroform and carbon tetrachloride were used. The p.m.r.
spectrum of lecithin was run in the presence and absence of the
particular compound under study. Changes in chemical shift
induced by the compound added indicated which protons of the
lecithin molecule were adjacent to, or possibly involved in, the
interaction. The involvement of the DDT or related compound
was established in a similar fasion by observing its p.m.r. spectra
as influenced by the addition of lecithin.

Binding of p,p'-DDT and Lecithin

Using this experimental approach the binding of p,p'-DDT to
lecithin has been established (11). The p.m.r. spectrum of β,γ-
dipalmitoyl-DL-phosphatidylcholine monohydrate in carbon tetrachloride (Fig. 1) shows resonance peaks with varying degrees of
multiplicity that correspond to protons "a" to "i". The chemical
shift of the resonance peaks of the water of hydration protons
varies with lecithin concentration and temperature as a consequence

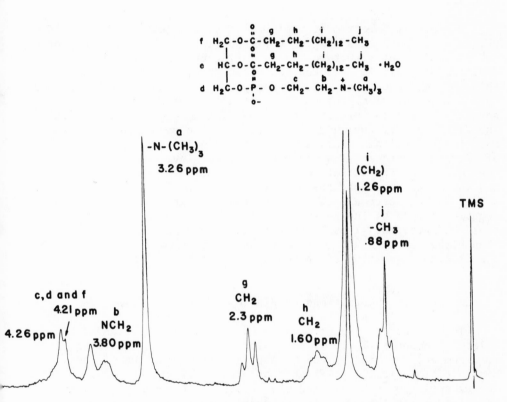

Fig. 1.   Proton magnetic resonance spectrum of DL-α lecithin in
          CHCl$_3$ at 34°C.   (Haque, Tinsley and Schmedding, J.
          Biol. Chem., <u>247</u>, 161, 1972).

of micelle formation (14). The addition of p,p'-DDT induced low-field changes in the chemical shift of resonance peaks due to protons "a", "b", and "c". The maximum change (11 Hz) was observed in the chemical shift of the resonance peak of the $-\overset{+}{N}(CH_3)_3$ protons indicating that these protons experienced a substantial change in their electronic environment in the binding process.

The p.m.r. spectrum of p,p'-DDT in carbon tetrachloride (Fig. 2) consists of a singlet corresponding to the benzylic proton "a" and multiplet structures for the ring protons "b" and "c". Addition of lecithin produced low-field changes in the chemical shift of resonance peaks of the benzylic proton"a" and ring protons "b", but not of ring protons "c". It would appear that the benzylic proton is involved in the binding of p,p'-DDT to lecithin. The fact that DDE, which molecule contains no such proton, does not bind to lecithin would support this conclusion. The proximity of the ring protons "b" would explain their involvement. As a consequence of the inductive effects of the chlorine substituents and the aromatic rings, the benzylic proton would be somewhat acidic in character and it is suggested that it would bind to the negative oxygen of the phosphorylcholine moiety. A hydrogen bond could be involved. That the $-\overset{+}{N}(CH_3)_3$ protons are more involved than either protons "b" or "c" of the lecithin may be explained by a cyclic configuration of the phosphorylcholine moiety.

Comparative Observations of Binding of Related Compounds:

These studies have been extended to include a series of DDT-related compounds (Table 1) to explore further the structural requirements for binding and to obtain some indication of the possible biological significance of such interactions (12). With the exception of DDE, the addition of these compounds to a dilute solution of lecithin resulted in changes in the chemical shift of lecithin proton resonance peaks "a", "b", and "c". As with DDT, the major change was observed with the $-\overset{+}{N}(CH_3)_3$ protons, again indicating the involvement of the phosphorylcholine moiety in the binding.

The addition of lecithin to dilute solutions of these compounds induced low-field changes in the chemical shift of the resonance peaks of the protons indicated (Fig. 3). The magnitude of these changes varied considerably among the compounds tested. It should be noted that any changes in the chemical shift of the benzylic proton peaks of the DDA and the two DDD isomers could not be observed due to interference of the lecithin spectrum. For compounds substituted in the para,para' position, only the protons

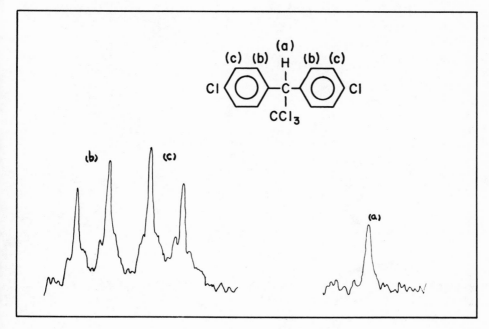

Fig. 2.   Proton magnetic resonance spectrum of p,p'-DDT in CCl₄.

Table 1.

DDT-type Compounds Used in the Studies

Compound	Common Name	Chemical Structure
1,1,1-Trichloro- ethanes	p,p'-DDT	$Cl\text{—}C_6H_4\text{—}CH(CCl_3)\text{—}C_6H_4\text{—}Cl$
o,p-DDT	o,p-DDT	$Cl\text{—}C_6H_4\text{—}CH(CCl_3)\text{—}C_6H_4(Cl)$
Methoxychlor	Methoxychlor	$CH_3O\text{—}C_6H_4\text{—}CH(CCl_3)\text{—}C_6H_4\text{—}OCH_3$
1,1-Dichloro- ethanes	p,p'-DDD	$Cl\text{—}C_6H_4\text{—}CH(HCCl_2)\text{—}C_6H_4\text{—}Cl$
o,p-DDD	o,p-DDD	$Cl\text{—}C_6H_4\text{—}CH(HCCl_2)\text{—}C_6H_4(Cl)$
1,1-Dichloro ethene	DDE	$Cl\text{—}C_6H_4\text{—}C(CCl_2)\text{—}C_6H_4\text{—}Cl$
1,1,1-Trichloro ethanol	Dicofol (Kelthane)	$Cl\text{—}C_6H_4\text{—}C(OH)(CCl_3)\text{—}C_6H_4\text{—}Cl$
Acetic acid	DDA	$Cl\text{—}C_6H_4\text{—}CH(COOH)\text{—}C_6H_4\text{—}Cl$

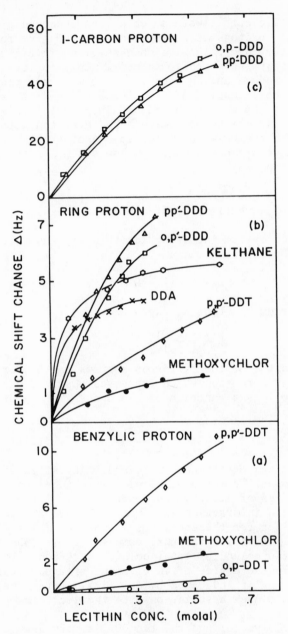

Fig. 3. Changes in chemical shift of various protons of DDT-type compounds with addition of lecithin. The concentrations of DDT-type compounds: p,p'-and o,p-DDT, 1.4×10^{-2} molal; methoxychlor, 1.49×10^{-2} molal; p,p'-DDD, 2.2×10^{-2} molal; o,p'-DDD, 1.5×10^{-2} molal; p,p'-DDA, 1.78×10^{-2} molal; dicofol, 1.39×10^{-2} molal. (Haque, Tinsley and Schmedding, Mol. Pharm. 9, 17, 1973).

ortho to the benzylic carbon were affected, as observed previously
with DDT. The ring protons of o,p-DDT and o,p-DDD gave more complex spectra and the chemical shift changes given in Fig. 3b are
for the one prominent sharp peak obtained in this region.
Lecithin did not produce any changes in the p.m.r. spectrum of
DDE. Thus, with the exception of DDE, the binding of these
compounds is similar to that observed with p,p'-DDT involving
the phosphorylcholine moiety of lecithin and a proton with
acidic character in the DDT-type molecule. Another class of
compounds, the vinyl organophosphates, show a similar type of
binding relationship (15).

Quantitative Treatment:

An equilibrium system involving lecithin (L) and the DDT
compound (D) may be represented:

$$L + D \underset{\rightarrow}{\overset{\leftarrow}{}} L \cdot D \qquad\qquad (i)$$

giving an equilibrium constant K, expressed:

$$K = \frac{[L \cdot D]}{[L]\,[D]} \qquad\qquad (ii)$$

Equilibrium constants for the different systems can be derived
from the following expression which expresses observed changes in
chemical shift ($\Delta\delta_o$) as a function of the concentration of added
lecithin C:

$$\frac{1}{\Delta\delta_o} = \frac{1}{K\Delta\delta_c} \cdot \frac{1}{C} + \frac{1}{\Delta\delta_c} \qquad\qquad (iii)$$

Using dilute solutions of the compounds under investigation,
changes in chemical shift for a particular resonance peak are
obtained for a series of lecithin concentrations. A typical plot
for the DDT-lecithin interaction is given in Figure 4. Regression
analysis gives values for the equilibrium constant (K) and the
change in chemical shift for the pure complex ($\Delta\delta_c$). The overall
tendency for the two molecules to associate in carbon tetrachloride
is given by K while the degree of involvement of a particular
proton in the binding is indicated by $\Delta\delta_c$ (12). Larger values for
this constant mean closer association with the lecithin.

p,p'-DDT, Methoxychlor and the two DDD isomers gave equilibrium constant of about the same order of magnitude (Table 2).
The introduction of a hydroxyl and carboxyl substituent in the
ehtane moiety (dicofol and DDA) increased the equilibrium
constant considerably. DDE did not bind to lecithin while o,p-DDT
interacted only to a limited degree. For a given system K values

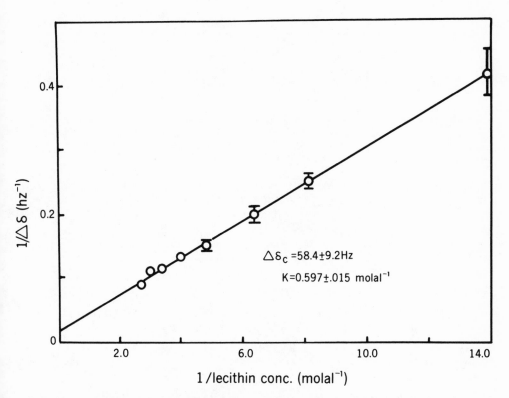

Fig. 4.   Reciprocal plot of change of chemical shift of benzylic
proton of DDT as a function of lecithin concentration at
33°C.   (Tinsley, Haque and Schmedding, Science, 174, 145,
1971).

Table 2.

Equilibrium Constants and Changes in Chemical Shift for the
Lecithin Complexes of DDT-type Compounds in CCl_4

Compound	Proton Studies	Equilibrium Constant, K Molal^{-1}	Chemical Shift Change, $\Delta\delta$ Hz
p,p'-DDT	Benzylic	0.598±0.16[a]	58.3±9.2[a]
	Ring	0.716±0.30	18.8±5.6
o,p-DDT		Changes in chemical shift too small for calculation	
Methoxychlor	Benzylic	2.52±0.28	5.44±1.42
	Ring	2.31±0.46	4.31±1.89
p,p'-DDD	C-1	2.21±0.04	104.0±3.7
	Ring	3.69±0.04	12.6±0.4
o,p-DDD	C-1	1.47±0.08	151.2±12.2
DDE		No interaction observed	
Dicofol	Ring	56.4±0.4	5.7±0.2
DDA	Ring	29.2±0.1	4.66±0.10

[a] Derived from error terms for slope and intercept of regression
equation.

agreed quite closely when derived from different proton resonance peaks. The value for $\Delta\delta_c$ varied over a wide range with large differences being observed between different protons on the same molecule and between the same proton on different molecules.

Biological Implications:

It is clear that these compounds can bind to lecithin and the logical question to ask is whether such interactions could occur in the cell. If so, DDT-type compounds would have the potential of affecting membrane structure and function. Although such compounds partition into hydrophobic regions of the cell, the degree to which the carbon tetrachloride environment could be reproduced is not known. Thus, the best that can be done at this stage is the development of circumstantial evidence by considering the structural features of the compounds which are important in lecithin binding in relation to those structural features known to be important for biological activity -- in this case, insecticidal activity.

The major evidence involves the role of the benzylic proton. Insecticidal activity is lost if the benzylic proton of p,p'-DDT is replaced by a chlorine or methyl group. This response could involve either some stereochemical requirement or a binding site. The high $\Delta\delta_c$ value indicates that the benzylic proton is a major factor in the binding of p,p'-DDT to lecithin. No binding or only limited binding results if this proton is absent (DDE), or relatively unavailable for steric reasons (o,p-DDT). These two compounds also have low insecticidal activity.

The degree to which a given proton is involved in the binding to lecithin is dependent on its acidic character. This is illustrated by the difference in the $\Delta\delta_c$ values for the benzylic protons of Methyoxychlor and p,p'-DDT. The chlorine substituent has a stronger inductive effect than the methoxyl substituent, hence the difference in the character of the benzylic protons. Although these two compounds are sterically equivalent the p,p'-DDT is more active as an insecticide, suggesting that the acidic character of the benzylic proton may also be a factor in their biological action.

The highest $\Delta\delta_c$ values were observed with the proton on the 1-carbon of DDD; a stronger inductive effect results when the chlorine substituents are on the same carbon as the proton. In cockroach nerve p,p'-DDD produced a response similar to, but weaker than that observed with p,p'-DDT (16). The location of the acidic or binding proton could thus be a factor.

The mode of action of these compounds doubtless involves a number of factors; however, the circumstantial evidence developed above would suggest that the tendency to bind to lipid could be one of these factors.

STUDIES IN AQUEOUS SYSTEMS

P.m.r. Spectra of DDT-Compounds in Liposomes and in Model Membranes:

Before extending these studies to aqueous systems it was necessary to establish procedures for dispersing the DDT-compounds in the aqueous phospholipid so that a clearly defined spectrum could be obtained. This was accomplished by first dissolving specified amounts (20-200 mg) of the phospholipid and DDT-compound (2-5 mg) in chloroform. The solvent was evaporated to dryness under vacuum and the sample dispersed in 1 ml of D_2O, using a Biosonik sonic oscillator (10 minutes at a setting of 30). It was observed that p,p'-DDT could be more readily dispersed using a phospholipid with a low transition temperature such as egg-yolk lecithin. The sample was tranferred to an n.m.r. tube and the spectrum recorded. Chemical shifts were reported with reference to an external tetramethylsilane lock standard.

Spectra which have been obtained using this procedure are given for the egg-yolk lecithin and the DDE and p,p'-DDT in the D_2O-lecithin dispersion (Fig. 5). The lines are broader and the spin-spin splitting of the ring protons is poorly resolved. The line broadening could result from increased viscosity, binding or inhomogeneity of the dispersion. Unfortunately, considering earlier studies, the resonance peak of the benzylic proton was masked by a large HDO peak.

A comparison has been made of the resonance peaks of the ring protons, considering chemical shift values observed in carbon tetrachloride and in the liposome system (Talbe 3). With the exception of DDE, a low-field change of about 7 Hz was observed for the ring protons "b" - protons ortho to the benzyl carbon. The ring protons "c" were virtually unaffected. These observations parallel the response of the ring protons observed in the non-aqueous system. With DDE again being atypical one could hypothe-size that the benzylic proton was again involved in some inter-action with the phospholipid. Further studies would be needed to validate this hypothesis.

The majority of the phospholipid would assume a liposome con-figuration and the DDT-compounds would be expected to be present

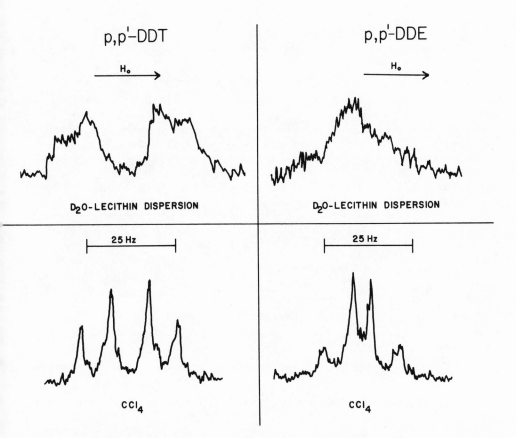

Fig. 5.   Proton magnetic resonance spectrum of DDT and DDE in
           CCl_4 and in ultrasonicated lecithin dispersions.

Table 3.

Chemical Shift Changes in the Ring Protons of DDT-type
Compound in CCl_4 and in Aqueous-Lecithin Dispersion

| Compound | Chemical Shift Change**(Hz) | | | | | |
| | Protons (b) | | | Protons (c) | | |
	CCl_4	Dispersion	Change	CCl_4	Dispersion	Change
p,p'-DDT	790.6	798.4	-7.8	771.0	770.5	0.5
p,p'-DDD	766.9	772.0	-5.1			
p,p'-DDA*	766.3	773.3	-7.0			
p,p'-DDE		No changes observed				
Dicofol	803.7	811.3	-7.6	768.3	770.8	-2.5

* Only one peak was observed in the dispersion. The center peak
   is compared.

** Chemical shifts are with reference to TMS as an external stand-
   ard.

within the bilayer forming these structures. Changes in the p.m.r.
spectrum of the lecithin were only observed upon incorporation of
DDA. Significant broadening of the resonance peaks of the methyl
and methylene protons of the fatty acid chains was observed (Fig.
6, Table 4). This would suggest that DDA produced restriction
in the molecular motion of the fatty acid chains, a response
comparable to that observed with cholesterol (17). DDA is more
soluble than the other compounds studied and the carboxyl is considerably more polar than other functional groups in the molecules
tested. Thus, DDA might be expected to be oriented at the lipid-
water interface rather than the interior of the bilayer.

Phospholipid Configuration and the p.m.r. Spectrum of DDE:

    We have observed that the p.m.r. spectrum of DDE dispersed
in D_2O-dipalmitoyl lecithin is influenced both by the concentration
of the lecithin and the temperature. These changes have been
interpreted with reference to changes in the configuration of
phospholipid (13).

    In a D_2O dispersion of dipalmitoyl lecithin (20 mg/ml) the
ring protons of DDE gave only the one resonance peak ($\delta\sim6.82$ p.p.m.).
With increasing concentration of lecithin a low-field peak
($\delta\sim7.16$ p.p.m.) appeared which increased in intensity with a concomitant decrease in intensity of the high-field peak (Fig. 7).
Using a concentration of lecithin which gave both peaks, it was
observed that increasing temperature favored the low-field peak
while the high-field peak disappeared at temperatures approximating
the transition temperature of the lipid (Fig. 7). Egg-yolk lecithin, which has a low transition temperature, gave only the low-
field peak.

    Two different electronic environments for the DDE are indi-
cated and the rate of exchange of DDE protons between the two
environments is slow enough that two peaks are observable. The
most feasible interpretation of these data involved a consideration of the various lecithin configurations attainable in the
aqueous dispersion. It has been suggested that the low-field peak
corresponds to the situation where the DDE is freely dispersed in
the hydrophobic "liquid" environment of the bilayer. The high-
field peak would correspond to an aggregated form of the DDE which
process occurs when there is insufficient "liquid phase" to disperse the DDE, i.e. low temperatures and low lecithin concentrations. The aggregated form is possibly included in the gel phase
of the lecithin. These data would indicate that some of the lecithin can exist in a liquid-crystalline phase below the transition
temperature of the pure lipid. This would be possible if the in-

Table 4.

Line Width Changes in the $(CH_2)_n$, CH_3 and Choline Protons of
Lecithin in Presence of DDA

Conc. of DDA mg/ml	Conc. of Lecithin mg/ml	Line Width (Hz)		
		$-(CH_2)_n$	$-CH_3$	$N^+(CH_3)_3$
0	40	22	20	6
5	40	52	25	10
5	60	38	21	9
5	80	36	21	8

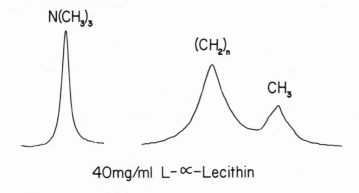

40mg/ml L-∝-Lecithin

←—50 Hz—→

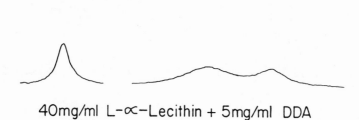

40mg/ml L-∝-Lecithin + 5mg/ml DDA

Fig. 6. Proton magnetic resonance spectrum of egg-yolk lecithin in presence and in the absence of DDA in D_2O ultrasonicated dispersion.

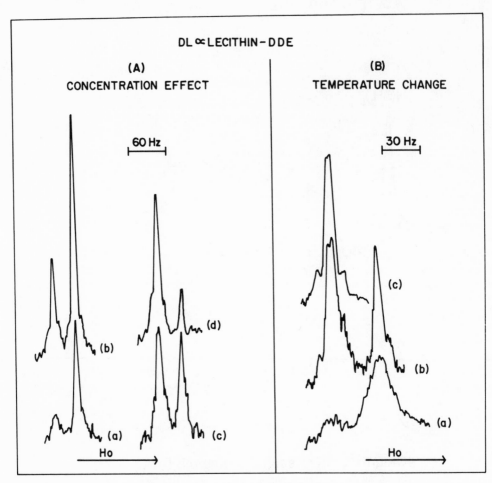

Fig. 7. (A) N.M.R. Spectra of DDE (5 mg/ml) ring protons in D_2O
dispersion in presence of varying concentration
(mg/ml) of DL-α-lecithin: (a) 20 (b) 35 (c) 45
(d) 100, at room temperature. (B) N.M.R. spectra
of DDE (5 mg/ml) ring protons in DL-α-lecithin
(45 mg/ml) D_2O dispersion at various temperatures:
(a) 20°C (b) 35°C (c) 44°C. (Haque, Tinsley and
Randall, Biochemistry, 12, 2346, 1973).

corporation of DDE broadened the temperature range over which the gel to liquid transition occurred.

The distribution of the DDE between the two environments can be expressed:

$$(DDE)_1 \underset{\leftarrow}{\overset{\rightarrow}{}} (DDE)_2 \qquad\qquad (iv)$$

and if C_1 and C_2 are the respective concentrations of DDE in the two environments, an expression for a partition coefficient K can be given

$$K = \frac{I_2}{I_1} = \frac{C_2}{C_1} \qquad\qquad (v)$$

I_2 and I_1 are the intensities of the low- and high-field peaks, respectively. Variation of K with temperature allows calculation of enthalpy changes.

These observations have been conducted with other types of complex lipids such as sphingomyelin, phosphatidyl serine and phosphatidyl ethanolamine. The K values as a function of lipid concentration are given in Table 5. Differences are observed in the chemical shift between the two peaks, the temperature required to eliminate the high-field peak and enthalpy changes. Some of these data are summarized in Table 6.

This study illustrated how these techniques can provide in- formation on the configuration of phospholipids. Another in- teresting aspect of the study was the fact that DDE showed a markedly different response to p,p'-DDT. Only one resonance peak (low-field) could be obtained with the p,p'-DDT. A high-field resonance peak was not observable. The greater amount of phospho- lipid required to disperse the DDT in the D_2O may also favor an environment corresponding to the low-field peak. Another possi- bility may be that DDE may aggregate more readily than the DDT.

CONCLUSIONS

1. DDT-type compounds will bind to lecithin in carbon tetrachlo- ride solution. P.m.r. procedures used to demonstrate this inter- action also provide information on thermodynamic parameters as well as structural aspects of the binding process.

2. The p.m.r. spectra of DDT-type compounds can be observed when these compounds are dispersed in liposome systems. Interaction data observed in such systems will be of greater biological sig-

Table 5.

Partition Coefficient K as a Function of Phospholipid Concentration

DL-α Lecithin Concentration mg/ml	K	Sphingomyelin Concentration mg/ml	K	Phosphatidyl Serine Concentration mg/ml	K
20	.3	25	0	20	.263
35	.57	50	.25	25	.533
45	1.02	100	1.04	35	1.09
100	6.4	150	1.30	50	--
		200	5.03		

Table 6.

Chemical Shift Difference Δδ Between the Two DDE Peaks, Transition
Temperature for Disappearance of High-field Peak and Enthalpy
Changes ΔH for Various Phospholipids

Phospholipid	Transition Temperature	Δδ Hz	ΔH K·Cal·mole^{-1}
DL α-Lecithin	~44°C	35 ± 2	4·8
Egg-yolk Lecithin	<25°C	----	---
Sphingomyelin	~80°C	29 ± 3	15·9
Phosphatidyl Serine	~80°C	17 ± 2	22·8; 1·2
Phosphatidyl Ethanolamine	~70°C	21	29·6

nificance than that observed in non-aqueous systems since the configuration of the phospholipids in the liposome is similar to the configuration in the natural membrane.

3. Comparisons can be made between the binding characteristics of these compounds and the biological response they produce. In particular, the distinct toxicological differences (toxicity, metabolism, etc.) between p,p'-DDT and DDE correspond to pronounced differences in the way these two compounds behave in the phospholipid systems studied.

4. Investigations of the p.m.r. spectra of DDE in liposomes can provide information on the configuration of the phospholipid in the liposome.

5. P.m.r. spectroscopy can provide information which will contribute to the understanding of the molecular basis of the mode of action of DDT-type compounds.

ACKNOWLEDGEMENTS

We thank Mr. David Schmedding for his technical assistance during the progress of this work. The work described here has been supported by the National Institute of Environmental Health grants No. ES-00210 and ES-00040.

REFERENCES

1. R. D. O'Brien, Insecticide Action and Metabolism, Academic Press, N.Y. (1967).

2. T. Narahashi and H. G. Hass, J. Gen. Physiol., 51, 177 (1968).

3. B. Hille, J. Gen. Physiol., 51, 191 (1968).

4. B. D. Hilton and R. D. O'Brien, Science, 168, 841 (1970).

5. G. Holan, Nature, 221, 1025 (1969).

6. G. Holan, Nature New Biol., 232, 644 (1971).

7. R. D. O'Brien and F. Matsumura, Science, 146, 657 (1964)

8. W. E. Wilson, L. Fishbein, S. T. Clements, Science, 171, 180, (1971).

9. R. T. Ross and F. J. Biros, Biochem. Biophys. Res. Comm., 39, 723 (1970).

10. N. K. Wilson, J. Amer. Cham. Soc., 94, 2431 (1972).

11. I. J. Tinsley, R. Haque and D. Schmedding, Science, 174, 145 (1971).

12. R. Haque, I. J. Tinsley and D. Schmedding, Mol. Pharm., 9, 17 (1973).

13. R. Haque, I. J. Tinsley and S. E. Randall, Biochemistry, 12, 2346 (1973).

14. R. Haque, I. J. Tinsley and D. Schmedding, J. Biol. Chem. 247, 157 (1972).

15. R. Haque, I. J. Tinsley and J. K. McCrady, Pest. Biochem. Physiol., 3, 73 (1973).

16. D. I. V. Lalonde and A. W. A. Brown, Can. J. Zool., 32. 74 (1954).

17. D. Chapman and S. A. Penkett, Nature, 211, 1304 (1966).

A STUDY OF INTERMOLECULAR COMPLEXES OF BIS (p-CHLOROPHENYL) ACETIC ACID AND SOME BIOLOGICALLY SIGNIFICANT COMPOUNDS

Ralph T. Ross and Francis J. Biros

Environmental Protection Agency, Primate and Pesticides

Effects Laboratory, P. O. Box 490, Perrine, Florida   33157

The mode of action of chlorinated hydrocarbons has received much recent attention, and several elegant studies have attempted to relate binding and complexation phenomena with toxic effects.[1-7] The occurrence and persistence of the compound 1,1,1-trichloro-2,2-bis (p-chlorophenyl) ethane, p,p'-DDT (I),in the environment, in addition

(I)

to the fact that its biochemical mode of action has not been unequivocally established, has resulted in a continued effort to determine its molecular interaction capabilities. To date, the study of complexation of p,p'-DDT with biological compounds has been limited due to its insolubility in aqueous systems. However, attempts have been made to establish, on a molecular level, the behavior of p,p'-DDT complexes in nonaqueous media, employing high resolution nuclear magnetic resonance (NMR) spectroscopy.[5,8,9] The use of the NMR technique to study the behavior of such complexes permits more precise characterization of the possible molecular associations involved.

An important metabolite of p,p'-DDT is bis (p-chlorophenyl) acetic acid (p,p'-DDA), II. Since the specificity of biologically

(II)

active materials lies in the structure of the complexes they form
with macromolecular receptors[10], any qualitative data on the chemistry and function of agent-receptor complexes yields insight into the
pharmacodynamic behavior of the agent molecule. Various chemical
and physical techniques including spectrophotometry, microcalorimetry,
chemical kinetics, structure-activity correlations involving chemical
modifications of either the agent or the receptor, and electrophoresis
have been used to establish specific molecular interactions of
biochemical importance. A significant limitation of these methods
is their inability to reflect relevant structural detail inasmuch as
they measure large alterations in chemical and physical properties.
Suitable models for this study are the complexes that p,p'-DDA may
form with both bovine serum albumin (BSA) and human serum albumin
(HSA). The vascular transport of p,p'-DDA may be an important step
in the excretion of p,p'-DDT from the liver, the principal metabolic
site.[11] The solubility of p,p'-DDA in slightly alkaline aqueous
solutions and the ease of interpreting the NMR spectrum make this
compound unique for the study of complexation with biological compounds. This study was made to directly test the existence of binding of p,p'-DDA with BSA and HSA and to examine the structure of the
complexes.

EXPERIMENTAL

Proton NMR spectra were obtained on a Varian Associates HA-100
high resolution NMR spectrometer operating at 100 MHz. Chemical
shifts were measured by a Hewlett Packard Model 5512A electronic
frequency counter which displays the pen position in Hz from the
lock signal, in this case an external capillary of tetramethylsilane.
Temperature control and measurements were made with a Varian Associates V-4333 variable temperature probe and a V-4343 temperature
controller.

All solutions for NMR studies were prepared with D_2O as solvent
and p,p'-DDA concentrations are expressed in moles/liter ($\underline{M}$). BSA
and HSA concentrations are in percent w/v i.e., in grams per 100 ml
final volume.

Except for the variable temperature studies, the NMR measurements were obtained at 31.5 ±1.0°C. The spin-spin relaxation times,

T_2, were obtained from a measurement of line width.[12] Each reported value of T_2 represents the mean of at least 6 measurements.
All pD adjustments were made with DCl-D_2O and NaOD-D_2O solutions.
pD measurements were made with a Beckman Expandomatic pH meter and
are uncorrected for deuterium isotope effects.

Commercial sources used were as follows: D_2O, tetramethylsilane, and NaOD, Wilmad Chemical Co., Buena, N. J.; p,p'-DDA,
Aldrich Chemical Co., Milwaukee, Wis.; crystallized bovine serum
albumin and human serum albumin, Sigma Chemical Co., St. Louis, Mo.

RESULTS AND DISCUSSION

The NMR spectrum of 0.25 M solution of p,p'-DDA is shown in
Figure 1. The aromatic protons are easily identified by their
characteristic low field position at 7.54 ppm. The methine proton
on the α carbon atom gives rise to a signal at 5.34 ppm. The remaining peak at 5.23 ppm is due to a water resonance arising from
HDO generated by the replaceable hydrogen atoms of the p,p'-DDA,
BSA, and HSA molecules.

To study the effects of BSA and HSA on the p,p'-DDA molecule,
0.25 and 0.40 M solutions of p,p'-DDA containing 0, 1.25, 2.50 and
4.00% of BSA and HSA respectively were examined. Figure 2 shows
the results of a typical series of measurements. The relaxation
rates of the aromatic protons are changed by a larger factor (x5)
than the relaxation rate of the α-methine proton (x2). Thus the
binding phenomenon may be interpreted in terms of preferential
stabilization of the aromatic rings by binding sites on the BSA
and HSA molecules.

Since the observed broadening may be nonspecific and may be due
to alternate relaxation mechanisms such as intermolecular interactions
between p,p'-DDA molecules, control experiments were performed on a
series of solutions at varying concentrations of p,p'-DDA. Chemical
shifts decreased as the concentration of p,p'-DDA increased, with a
maximum shift of 18.2 Hz upfield for the aromatic resonances between 0.05 M and 2.00 M solutions. Differences in relaxation rates
comparable to those observed on addition of BSA, however, were not
observed.

The large upfield shift of the aromatic proton resonances may
be attributable to extensive association of p,p'-DDA molecules at
higher concentrations, resulting in an increase of shielding effects
above and below the aromatic rings.[13] Further, an apparent collapse
of the AA'BB' multiplet[14] of the two sets of aromatic protons to a
single resonance line occurred at concentrations exceeding 0.20 M
p,p'-DDA, under our conditions. Thus, specific molecular interactions involving solute or solvent molecules preferentially influence either the AA' or the BB' protons of the aromatic rings of

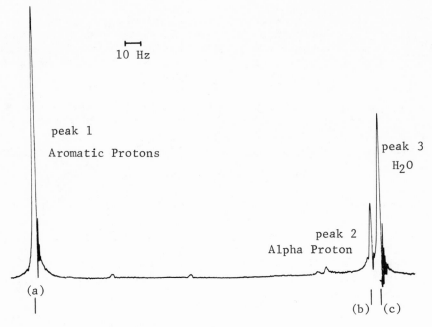

Figure 1

The NMR spectrum of 0.25 $\underline{M}$ solution of p,p'-DDA
(a) Aromatic protons (753.7 Hz)
(b) Benzylic proton   (533.7 Hz)
(c) Water protons (523.4 Hz)

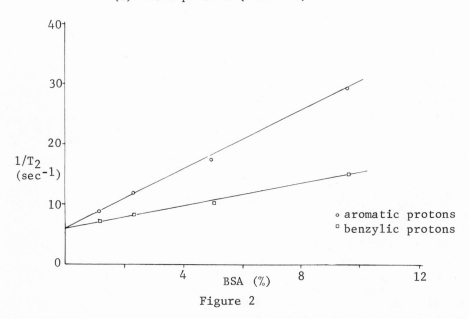

Figure 2

Plots of $1/T_2$ of a 0.25 $\underline{M}$ solution of p,p'-DDA vs BSA concentration.

Table 1. Effects of Ionic Strength on Chemical Shifts (Hz)
          and $1/T_2$ (Sec^{-1}) Values of p,p'-DDA Spectral Lines

NaCl Conc. M	Aromatic Protons			Tertiary Protons		
	Chemical Shift (Hz)	$1/T_2$	(Sec^{-1})	Chemical Shift (Hz)	$1/T_2$	(Sec^{-1})
A. 0.25 M p,p'-DDA, 5% Albumin						
0.0	761.2	9.97	(0.18)	531.7	8.82	(0.53)
0.1	760.5	16.70	(0.42)	532.5	14.85	(0.16)
0.3	760.8	21.62	(0.22)	532.1	17.61	(0.09)
0.7	759.9	27.15	(0.27)	532.0	19.18	(0.23)
1.0	759.3	31.11	(0.5)	531.7	20.75	(0.31)
2.0	757.3	64.05	(0.31)	533.7	37.81	(0.11)
B. 0.25 M p,p'-DDA						
0.0	763.5	5.28	(0.62)	533.2	4.99	(0.10)
0.1	762.4	5.49	(0.35)	532.6	5.37	(0.35)
0.3	761.1	5.31	(0.49)	532.7	5.59	(0.26)
0.7	760.4	5.21	(0.3)	532.4	5.65	(0.08)
1.0	760.0	5.87	(0.22)	533.8	5.87	(0.32)
2.0	759.8	6.15	(0.48)	532.8	6.09	(0.46)

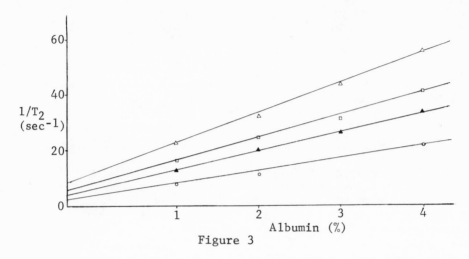

Figure 3

The effect of pD on the relaxation rates of 0.25 M
solutions of p,p'-DDA at varying albumin concentrations
(%). Δ p,p'-DDA·BSA pD 7.45; ☐ p,p'-DDA·BSA pD 8.25;
o p,p'-DDA·BSA pD 9.70; ▲ p,p'-DDA·HSA pD 7.70.

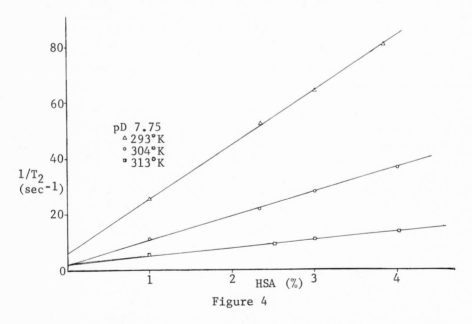

Figure 4

Temperature dependence of relaxation time vs HSA concentration.

p,p'-DDA at higher concentrations.[15],[16] This effect facilitated relaxation measurements of the aromatic ring protons.

Control experiments in which the chemical shifts of the AA' protons and the BB' protons were measured as a function of increasing BSA concentration have shown that virtually no contribution to the broadening of the aromatic proton resonances can be attributed to small internal chemical shift variations and incipient resolution of the complex, collapsed AA'BB' multiplet.

To study the effects of pD on the p,p'-DDA interactions with BSA and HSA, solutions of 0.25 $\underline{M}$ p,p'-DDA containing 0, 1.0, 2.5, 3.0, and 5.0% BSA and HSA were examined over the pD range 7.45-9.75. The measured relaxation rates were quite dependent on pD, as expected from a specific binding mechanism (Figure 3); the largest values are at pD 7.45 and decrease as the pD is increased, a finding consistent with the known behavior of albumin binding phenomena. BSA is known not to bind above pD 10.5 because of protein denaturation.[17]

The effect of ionic strength on the p,p'-DDA albumin interaction was studied with a solution of 0.25 $\underline{M}$ p,p'-DDA plus 5% BSA, together with a control solution of 0.25 $\underline{M}$ p,p'-DDA, in the presence of varying amounts of 0.0-2.0 $\underline{M}$ NaCl. The results of the line width measurements are given in Table 1. The increase in line width with increasing ionic strength is pronounced for p,p'-DDA·BSA but small for p,p'-DDA alone; the chemical shifts of the peaks of both solutions change in the same direction, upfield with increasing NaCl concentration.

These line width measurements are useful in deciding on a mechanism to explain p,p'-DDA·BSA binding. Since the energy of electrostatic attractions varies inversely with dielectric constant,[18] one would expect an ionic bond between p,p'-DDA and BSA to be weakened by addition of salt. The reverse of this situation is observed. These findings suggest that p,p'-DDA·BSA interaction is hydrophobic, as expected, since the aromatic rings are the primary site of binding as discussed above.

The temperature dependence of the NMR spectra of p,p'-DDA with HSA is illustrated in Figure 4. Line widths for both p,p'-DDA·BSA and p,p'-DDA·HSA complexes were found to increase as the temperature decreased. This effect is probably due to an increase in the fraction of p,p'-DDA bound as the temperature decreases, since changes in line widths for unbound p,p'-DDA are not sufficient to account for the marked increase in relaxation rates.

To test the specificity of the binding sites on the albumin molecules, a series of inhibition studies was made. In these experiments an attempt was made to displace the p,p'-DDA from the

Table 2. Inhibition Studies with p-Chlorophenylacetic Acid; $1/T_2$ Values (Sec^{-1})

Sample Description	p,p'-DDA Spectral Lines		p-CPA Spectral Lines	
	Aromatic Protons	-CH-	Aromatic Protons	-CH-
0.25 M p,p'-DDA	3.11 (0.16)	2.97 (0.31)	------------	--------
1.20 M p-CPA	------------	-----------	3.45 (0.19)	3.19 (0.25)
0.25 M p,p'-DDA 1.20 M p-CPA	OVERLAP	3.79 (0.42)	OVERLAP	3.57 (0.11
0.25 M p,p'-DDA 5% BSA	------------	-----------	10.24 (0.48)	5.33 (0.35)
0.25 M p,p'-DDA 1.20 M p-CPA 5% BSA	8.86[a] (0.45)	4.97 (0.22)	9.18[a] (0.58)	4.45 (0.18)

[a]Slight overlapping of peaks occurred; these values are estimates.

albumin by adding a large excess of another chemically analogous compound, p-chlorophenyl acetic acid (p-CPA).

In the NMR spectrum of p-CPA the methylene resonances are dis-
placed upfield to 390.5 Hz relative to the tertiary proton in p,p'-
DDA. Thus the relaxation rates of both these peaks in a mixture of
these two compounds can be measured. Unfortunately, the aromatic
proton resonances overlap somewhat and measurements of the relaxation
rates of these peaks, where given, represent estimates. Table 2
gives the line widths of the resonance peaks of interest. The control
experiments indicate that both p,p'-DDA and p-CPA bind to BSA.
Furthermore, the spectrum of the p,p'-DDA, p-CPA and BSA mixture
shows slightly less broadening of the p-CPA peaks and much less
broadening of the p,p'-DDA peaks compared with controls. This
difference may be due to a competition between p-CPA and p,p'-DDA
for the binding sites on the serum albumin molecule.

SUMMARY

The results of this investigation have shown that with the use
of high resolution NMR relaxation measurement studies, some insight
into the importance of molecular binding of p,p'-DDA with BSA and
HSA may be obtained. The binding phenomena may be interpreted in
terms of preferential stabilization of the aromatic rings by binding
sites on the BSA and HSA molecules. This was demonstrated by the
greater relaxation rates found for the aromatic protons of the p,p'-
DDA than those of the α-methine proton. These relaxation rates
were quite pD dependent as would be expected from a specific binding
mechanism. The binding of p,p'-DDA·BSA and p,p'-DDA·HSA is hydrophobic in nature since a direct relationship was found with ionic
strength and relaxation times. The relaxation rates of both complexes
increase with decrease in temperature.

BIBLIOGRAPHY

1. F. A. Gunther, R. Blinn, G. E. Carmen and R. L. Metcalf,
    Arch. Biochem. Biophys., 50, 504 (1954).

2. L. J. Mullins, Science, 122, 118 (1955).

3. F. Matsumura and R. D. O'Brien, J. Agr. Food Chem., 14, 36 (1966).

4. F. Matsumura and K. C. Patil, Science, 166, 212 (1969).

5. R. T. Ross and F. J. Biros, Biochem. Biophys. Research Commun.,
    39, 723 (1970).

6. N. K. Wilson, J. Amer. Chem. Soc., 94, 2431 (1972).

7. N. K. Wilson and W. E. Wilson, <u>The Science of the Total Environ.</u>, <u>1</u>, 245 (1972).

8. W. E. Wilson, L. Fishbein and S. T. Clements, <u>Science</u>, <u>171</u>, 180 (1971).

9. R. Haque, I. J. Tinsley and D. Schmedding, <u>Mol. Pharmacol</u>, <u>9</u>, 17 (1973).

10. O. Jardetsky and H. G. Wade-Jardetsky, <u>Mol. Pharmacol</u>, <u>1</u>, 214 (1965).

11. W. J. Hayes, Jr., "Annual Review of Pharmacology", Vol. 5, Annual Review Inc., Palo Alto, Calif., 1965, p. 27.

12. J. J. Fisher and O. Jardetsky, <u>J. Amer. Chem. Soc.</u>, <u>87</u>, 3237 (1965).

13. A. D. Broom, M. P. Schweiger and P.O.P.Ts'o, <u>J. Amer. Chem. Soc.</u>, <u>89</u>, 3612 (1967).

14. J. W. Eimsley, J. Feeney and L. H. Sutcliff, "High Resolution Nuclear Magnetic Resonance Spectroscopy," Vol. 1, Pergamon Press, London, 1965, p. 399.

15. T. Schaefer and W. G. Schneider, <u>J. Chem. Phys.</u>, <u>32</u>, 1218 (1960).

16. T. Schaefer and W. G. Schneider, <u>J. Chem. Phys.</u>, <u>32</u>, 1224 (1960).

17. C. Tanford, J. G. Buzzell, D. G. Rands and S. A. Swanson, <u>J. Amer. Chem. Soc.</u>, <u>77</u>, 6421 (1955).

18. A. A. Frost and R. G. Pearson, "Kinetics and Mechanism," 2nd Ed., John Wiley & Sons, New York, 1961, pp. 150-55.

NMR OF ADSORBED MOLECULES WITH A VIEW TOWARD PESTICIDE CHEMISTRY

H. A. Resing

Surface Chemistry Branch, Naval Research Laboratory

Washington, D. C. 20375

INTRODUCTION

We ask of pesticides that they kill pests and, having done that, bother us no more. Having neglected the second desideratum we are faced with the twin necessities of (a) designing new pesticides which meet this second requirement and (b) assaying and if possible, remedying the damage produced through this neglect. In this article we are concerned with the application of NMR methods to studies of adsorbed molecules, i.e. to studies of surface-chemical systems. Thus, we are in a sense two steps removed from a direct assault on the above problems, the first being the linking of pesticide chemistry to surface chemistry, and the second being a demonstration of the relevant insights gained from the application of NMR to surface chemistry. The first step is easy: a pesticide molecule upon application must adhere to some desired substrate; then it must chemically inter-act with the macromolecular structure of the pest and cause its destruction; it must next be eluted from the soil by means of natural waters: all of these processes depend on surface chemi-cal adsorption desorption equilibria, on the transport of mole-cules from point to point by diffusion or large scale chromato-graphic processes. The second step is also easy, because NMR line widths, shapes, relaxation times, and chemical shifts tell about the kinetics of chemical reactions, rates of molecular ro-tation and diffusion, the "chemical" environments of nuclei, and the orientation of molecules in space; this is true whether the state of matter under observation is liquid, solid, gaseous or "adsorbed". The application of NMR to surface science has been recently reviewed, and copiously (1,2,3,4). Topics briefly

discussed below are the mobility of interfacial water (3) and, as
a model pesticide, the mobility and chemical shift anisotropy of
adsorbed benzene (5). The reader must consult more general works
(6,7,8) for detailed definitions and experimental procedures.

THE MOBILITY OF INTERFACIAL WATER

From NMR relaxation times it is possible to deduce a correlation time τ, which is essentially the time between molecular
jumps (9). Because of hydrogen bond formation, rotational jumps
and diffusional jumps in water may be characterized by a single
jump time (9); this is in contrast, for instance, to non-
associated high symmetry molecules such as benzene, for which two
correlation times are necessary, one to characterize rotation
about the hexad axis (10), and the second to characterize the
jumps which reorient the hexad axis and translate the molecule.
However, to fit the observed relaxation time data for adsorbed
molecules, it has been necessary to allow this jump time to be
distributed over a broad range of values (11), so broad that it
is convenient to use a Gaussian distribution on a logarithmic
scale (the log-normal distribution); the data are then characterized by a width and a median jump time. These median jump times
for the systems water on charcoal (11) (pore diameter ~ 30Å),
water filling the pores (about 13Å diameter) of zeolite 13-X (12),
and water at 20% by weight on a bacterial polysaccharide membrane (13), are summarized in Fig. 1, where they are compared
with similar mobility estimates for supercooled bulk water and
for ice. From these studies we may conclude (a) that the
mobilities of these interfacial waters are much closer to that of
liquid water than to ice, and (b) that water in these micro-
regions does not readily freeze. The implications for pesticide
chemistry are more conceptual than practical; for instance in
setting up a kinetic picture for the interaction of a pesticide
molecule with receptor site in some insect, it is not necessary
to view either molecule as protected by some rigid sheath of ice
which must be "penetrated" for the desired interaction to occur.
Nor is it a viable hypothesis to propose that a pesticide molecule may be locked onto a soil particle by a rigid water sheath.

BENZENE ADSORBED ON CHARCOAL

Unfortunately this is the closest analogue to a pesticide
molecule yet studied by these NMR relaxation methods. For diffusion of a benzene molecule along the charcoal surface the data(5)
yield a median jump time of $\tau = 4.8 \times 10^{-11}$ sec at room temperature, which is of the same magnitude as the estimates for adsorbed
water in Fig. 1; the activation energy for this process is shown
to be 5.7 kcal/mole.

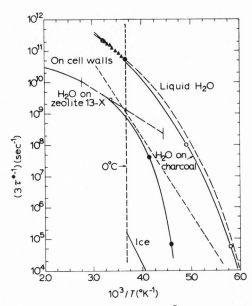

Fig. 1. Median jump frequency $(3\tau*)^{-1}$ versus inverse temperature for water, ice, and adsorbed water versus inverse temperature.

In all of the above systems relaxation times and/or absorption linewidths are dominated by the interaction of a nuclear magnet with its neighboring nuclear magnets, the "dipolar" interaction (9). This interaction in liquids is "averaged away" by rapid molecular motions, leaving only the chemical shift interactions which molecular motions do not average away; the result is the well known wealth of analytical benefits which high resolution NMR affords us (8). For adsorbed molecules, even though the average rate of molecular motion is fast, there is most often a "residual" dipolar interaction and resultant line broadening, due to molecular exchange with some few molecules bound to the surface in such a way that their jump times are much longer (4); then the analytical benefits and the subtle indications of molecular interactions obtainable by study of chemical shifts are for the most part lost. There are some recently devised techniques for instrumentally erasing nuclear dipolar contributions to the linewidth and revealing chemical shift tensors in the form of "powder" anisotropy patterns (14, 15). Thus, instead of a single sharp line which characterizes a chemical shift in the liquid, a broad spectrum results which is able to provide in many instances the three principal values of the chemical shift tensor, and provide an even more subtle indicator of perturbations of the molecular electronic structure. As an example the ^{13}C chemical shift anisotropy pattern at -195° C

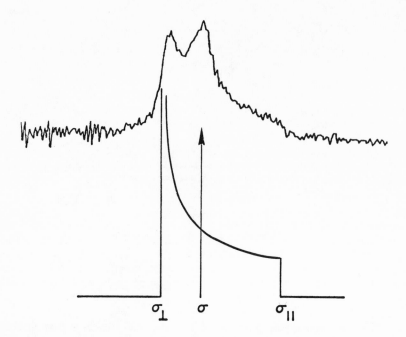

Fig. 2. ^{13}C chemical shift anisotropy powder patterns: (a) for
benzene adsorbed on charcoal, and (b) for solid benzene. $\sigma_\perp$ =
$(\sigma_{11} + \sigma_{22})/2$ is the average value in the plane of the benzene
ring; $\sigma_{\parallel}$ (= σ_{33}) is the principal value of the shift tensor
parallel to the symmetry axis; σ is the isotropic average.

for benzene adsorbed on charcoal (16) is shown in Fig. 2. This
spectrum is a superposition of two spectra, a narrow central peak
due to isotropically rotating molecules, and an "axially symmetric"
powder pattern due to molecules rotating about their respective
hexad axes only. The principal values of the chemical shift are,
within error, the same as those of pure solid benzene. The
implication here is that for some selected functional group of
a pesticide molecule, enriched in ^{13}C, electronic perturbations
due to some substrate (i.e. the details of the interaction)
might be revealed.

SUMMARY

For the most part the role of NMR relaxation studies of molecular motion in pesticide chemistry will be to provide background knowledge of the motional properties of adsorbed molecules. This is mainly for the simple reason that successful interpretation of the data requires simple systems. The "solid state" high resolution techniques on the other hand, especially those involving ^{13}C, offer promise of revealing the details of pesticide-substrate interactions.

REFERENCES

1. H. Pfeifer, in NMR, Basic Principles and Progress 7, 53 (1972)
2. E. G. Derouane, et al, in Advances in Catalysis (in Press).
3. H. A. Resing, Advances in Molecular Relaxation Processes 3, 199 (1972). This whole volume contains much of interest concerning macromolecules in solution, diffusion in biological media, and exchange kinetics, all via NMR.
4. H. A. Resing, ibid 1, 109 (1968).
5. J. K. Thompson, H. A. Resing and J. J. Krebs, J. Chem. Phys. 43, 3853 (1965).
6. E. R. Andrew, Nuclear Magnetic Resonance, Cambridge, 1955.
7. A. Abragam, The Principles of Nuclear Magnetism, Oxford, 1961.
8. J. A. Pople, W. G. Schneider, and H. J. Bernstein, High Resolution Nuclear Magnetic Resonance, McGraw-Hill, New York, 1959.
9. N. Bloembergen, E. M. Purcell, and R. V. Pound, Phys. Rev. 73, 1168 (1948).
10. E. R. Andrew and R. G. Eades, Proc. Roy. Soc. (London) A218, 537 (1953).
11. H. A. Resing, J. Chem. Phys. 43, 669 (1965).
12. H. A. Resing and J. K. Thompson, Adv. in Chemistry Series No. 101, American Chemical Society, 1971, p. 473.
13. H. A. Resing and R. A. Neihof, J. Colloid and Interface Sci. 34, 480 (1970).
14. M. Mehring, R. G. Griffin and J. S. Waugh, J. Chem. Phys. 55, 746 (1971).
15. A. Pines, M. G. Gibby, and J. S. Waugh, J. Chem. Phys. 56, 1776 (1972).
16. S. Kaplan, H. A. Resing and J. S. Waugh, submitted for publication.

NMR STUDIES OF PREFERENTIALLY ORIENTED WATER AT INTERFACES

D. E. Woessner

Mobil Research and Development Corporation

P. O. Box 900, Dallas, Texas 75221

INTRODUCTION

The hydrogen nuclear magnetic resonance (NMR) spectrum of bulk liquid water is a single, narrow line. This narrow singlet spectrum is observed because the molecules have rapid rotational and translational motions together with random orientations in the magnetic field of the spectrometer. However, the water molecule is somewhat electrically non-spherical and, as has been shown[1], instantaneous local order of molecular position and orientation does exist in the liquid. The rapid molecular motions result in a random average orientation if a given molecule is observed for even a short time interval such as 10^{-10} second.

On the other hand, a non-random average orientation of a water molecule can be induced by an immobile interface. The NMR spectrum of this preferentially oriented water can be a doublet. Such doublet spectra have been observed for water in contact with a number of different inorganic and organic materials including zeolites[2], montmorillonoid clays[3-7], collagen[8,10], concentrated soap pastes[11,12], lithium-DNA[13], keratin[14], rayon[15], Kelzan[16], and lyotropic liquid crystals[17,18].

The preferential orientation and the value of the doublet splitting should be influenced by molecular interactions of water at or near the interface. This paper reports doublet splittings which have been measured for a number of different clay-water systems in order to investigate the effects of substrate characteristics on these molecular interactions. Montmorillonite type clays[19] are convenient substrate materials since the fundamental particle is an aluminosilicate sheet approximately ten angstroms thick and hundreds of angstroms in length and width. These sheets are stacked parallel to form crystallites. It is possible to prepare oriented samples in which the crystallites are parallel. The addition of water to montmorillonoids results in expansion such that layers of water alternate with the parallel clay sheets. Also, different montmorillonoids can have different surface and electrical charge characteristics.

NMR DOUBLET SPLITTING AND PREFERENTIAL ORIENTATION

In order to understand the observed values of NMR doublet splitting of water molecules in interfacial systems, it is useful to examine the spectra which are expected from molecules in various states of motion. The nuclear spin energy level diagram[20] for the pair of protons in an isolated, motionless water molecule consists of three unequally spaced energies. Since only the transitions among neighboring energy levels are allowed, the NMR absorption spectrum is a doublet with the frequencies

$$\omega_a = \omega_o + \omega' \tag{1a}$$

$$\omega_b = \omega_o - \omega', \tag{1b}$$

where

$$\omega_o = \gamma H_o. \tag{2}$$

In Eq. (2), γ is the nuclear magnetogyric ratio and H_o is the strength of the magnetic field of the NMR spectrometer. The value of ω' depends on the structural parameters of the water molecule and on the orientation of the water molecule in the magnetic field. For both protons and deuterons, ω' may be expressed as

$$\omega' = B_o (3 \cos^2\theta - 1). \tag{3}$$

The angle θ refers to the angle between the direction of H_o and the line joining the two protons in the H_2O molecule. In the case of a deuteron in water, θ is the angle between the symmetry axis of the electrostatic field gradient at the deuteron, which is along the O-D bond, and the direction of the stationary magnetic field H_o. The constant B_o is different for the different hydrogen isotopes. For protons,

$$B_o = \frac{3}{4} \gamma^2 \hbar r^{-3}, \tag{4}$$

where $\hbar$ is Planck's constant divided by 2π and r is the distance between the two protons in H_2O. The value of B_o for a deuteron in water is

$$B_o = - \left(\frac{3}{8}\right) e^2 qQ\hbar^{-1}, \tag{5}$$

where $e^2qQ\hbar^{-1}$ is 2π times the deuteron electric quadrupole coupling constant in water.

It is convenient to use pulsed NMR techniques to study doublet splitting phenomena. In the pulsed NMR method, the nuclear spins are perturbed by the application of short, intense pulses of radio frequency energy. This perturbation produces a magnetization which has a rotating component in the plane perpendiuclar to the magnetic field of the NMR spectrometer. The rotating transverse magnetization induces an electric current in a coil of wire enclosing the sample. Therefore, the observable pulsed NMR signal is described by this transverse magnetization.

Each pulse used in the pulsed NMR experiment is identified by the number of angular degrees through which it rotates the nuclear spin magnetization. The expectation value of transverse magnetization following a 90° pulse is given by the expression[5]

$$\langle M_+(t) \rangle = iM_o \exp(-i\omega_o t) \cos(\omega' t). \qquad (6)$$

Following a 90°-180° pulse sequence,

$$\langle M_+(t) \rangle = -iM_o \exp[-i\omega_o(t-2\tau)] \cos(\omega' t). \qquad (7)$$

Following a 90° - 90° pulse sequence,

$$\langle M_+(t) \rangle = \tfrac{1}{2}iM_o \left\{ \exp(-i\omega_o t) \mp \exp[-i\omega_o(t-2\tau)] \right\}$$
$$\times \cos[\omega'(t-2\tau)]. \qquad (8)$$

In these equations, M_o is the equilibrium longitudinal nuclear spin magnetization parallel to the magnetic field, and τ is the time between the pulses in a given pulse sequence. The transverse magnetization rotates at the resonance frequency ω_o. The doublet splitting gives rise to the low frequency cosine terms which modulate the amplitude of this high frequency rotating magnetization.

Molecular rotational motions involving the angle θ cause ω' to be time-dependent so that Eqs. (6)-(8) no longer apply to the NMR signals from water. If the quantity $[3 \cos^2\theta(t)-1]$ is transformed into a local coordinate system, we obtain[5] the expression

$$[3 \cos^2\theta(t)-1] = 3(\ell \sin\theta' \sin\Psi'$$
$$+ m \sin\theta' \cos\Psi' + n \cos\theta')^2 - 1. \qquad (9)$$

In this equation, ℓ, m, and n are the time-dependent direction cosines of the nuclear interaction vector in the local coordinate system. The symbols θ' and Ψ' are two of the Euler angles (ϕ', θ', Ψ') which describe the orientation of the local coordinate system

in the coordinate system attached to H_o. Molecular motions cause
ℓ, m, and n to be time-dependent. Calculations have been made of
the pulsed NMR signals expected when the molecular reorientation is
described by jumps through the tetrahedral angle[21] and by a rotational diffusion[22] process. When the motions of a water molecule
are sufficiently rapid, Eqs. (1) and (3) become valid if ω' is replaced by its time-average value. In liquid water, all orientations
of the water molecule are equally probable and the average value of
$(3 \cos^2\theta-1)$ is zero. Consequently, the ω' value in liquid water is
zero and the NMR spectrum is a single line at frequency ω_o.

However, molecular interactions allow surfaces or interfaces to
restrict the possible orientations of the water molecules and cause
$\langle 3 \cos^2\theta-1 \rangle_{av}$ to be non-zero. When the correlation times τ_c which
describe the fluctuations of ℓ, m and n in the local coordinate
system are sufficiently short, the pulsed NMR signals are given by
Eqs. (6) to (8) in which the splitting ω' is replaced by its time-
average value. Numerical calculations [21] show that the relationship
$B_o \tau_c < 0.01$ describes the necessary condition for clays.

Since the presence of an interface makes it possible to observe
a doublet splitting due to preferential orientation of water, the
value of the splitting depends both on the orientation of the interface in the magnetic field of the NMR spectrometer and on the molecular preferential orientation with respect to the interface. If the
molecules at an assumed planar interface exchange positions rapidly,
and if the molecular orientation probability distribution is axially
symmetric about the normal to the interface, the effective value of
ω' is given by

$$\omega' = B (3 \cos^2\theta' -1), \tag{10}$$

where θ' is the angle between the direction of H_o and the normal to
the interface. The doublet splitting constant B for protons is
given by

$$B = \frac{3}{8} \gamma^2 \hbar r^{-3} \langle 3 \cos^2\theta''-1 \rangle, \tag{11}$$

where θ'' is the angle between the proton-proton vector and the
normal to the interface. For deuterons, the value of B is

$$B = - \frac{3}{16} e^2 qQ\hbar^{-1} \langle 3 \cos^2\theta'' -1 \rangle, \tag{12}$$

where θ'' is now the angle between the symmetry axis of the intramolecular electrostatic field gradient experienced by the deuteron
and the normal to the interface.

It is possible that only a fraction F of the water experiences
the interface while the remainder of the water behaves like bulk

liquid water. In this case, the normalized transverse relaxation curve is the following sum of the separate signals from the two types of water:

$$A = F \cos(\omega't) + (1-F). \tag{13}$$

DETERMINATION OF THE DOUBLET SPLITTING CONSTANT

The time dependence of the transverse nuclear spin magnetization can be evaluated accurately by measuring the spin-echo amplitude at time $t = 2\tau$. The curve measured by use of $90° - 180°$ pulse sequences is called a T_2 curve, while that obtained from $90° - 90°$ pulse sequences is called a T_3 curve. The times at which these curves reach $1/e$ of the initial value at time $t = 0$ are T_2^* and T_3^*, respectively. Equations (7) and (8) show that doublet splitting causes T_3^* to be much greater than T_2^*. In the absence of doublet splitting, $T_3^* = T_2^*$. Hence, measurements of both T_2 and T_3 curves provide a means of detecting doublet splitting.

As shown by Eqs. (10) - (12), the doublet splitting constant B characterizes the preferential orientation of water molecules by interfaces. In the case of oriented samples, the value of B can be found from the use of transverse T_2 relaxation curves measured with a known sample orientation. The observable cosine variations and the splitting constant are related by Eqs. (7) and (10). If the sample is non-uniform, it may contain a distribution of B values. The effect of such a distribution is to introduce a damping of the cosine variations.

Ordinarily, it is more convenient to prepare samples containing randomly oriented crystallites or "books" of parallel clay platelets than it is to prepare oriented samples. Such powder samples contain a distribution of ω' values ranging from $\omega' = 0$ to $\omega' = 2B'$. This distribution complicates the interpretation of T_2 transverse relaxation curves. A relationship between the T_2 curve and B is needed. As a first approximation which is accurate when water molecules are constrained to given "books", it is sufficient to substitute into Eqs. (6) and (7) the quantity $\cos(\omega't)$ averaged as follows over a random distribution of θ' values:

$$\langle \cos(\omega't) \rangle_{av} = \int_{0}^{\frac{1}{2}\pi} \cos[Bt(3\cos^2\theta'-1)] \sin\theta' \, d\theta'. \tag{14}$$

This function is shown in Fig. 1. One way to obtain the value of the splitting constant is to compare the observed maxima, minima, or zero-crossings to those of this computed curve. However, a distribution of B values tends to obscure these quantities. In particular, a very wide distribution can cause a smooth, bell-shaped decay curve when measurements using a single $90°$ pulse or $90°-180°$ pulse sequences are used. Such decay curves do not give any indication

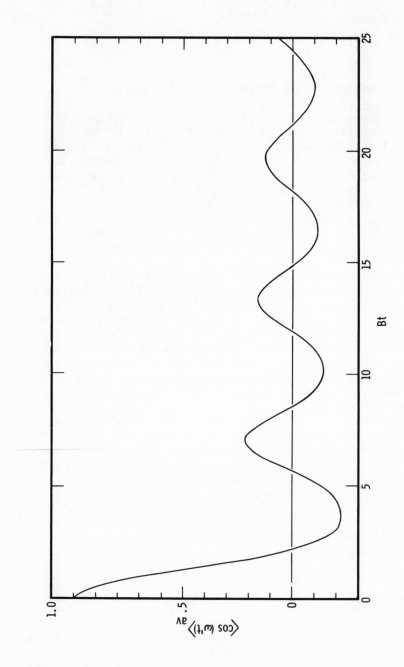

Figure 1.   The computed T_2 transverse relaxation curve for a powder sample.

of the presence of doublet splitting. Nevertheless, the additional measurement of the signals from 90°-90° pulse sequences can be used to detect doublet splitting.

An alternative method of using Eq. (14) can be used to evaluate B or the average B value. The quantity $\langle\cos(\omega't)\rangle_{av}$ decreases from unity at time $t = 0$ to $1/e$ at $t = 1.457/B$. Also, the curve $\langle\cos(\omega't)\rangle_{av}$ versus Bt is approximately linear in the range of Bt values between 0.8 and 1.8. Therefore, the value of B can be found from the relationship

$$B = 1.457/T_2^*.$$ (15)

This treatment of pulsed NMR signals assumes that ω' is time-independent. However, the translational diffusion causes water molecules to leave the interlaminar space in one crystallite and enter that of another. When the second crystallite has a different orientation in H_o, these water molecules experience a different θ' and consequently a different ω' value, as Eq. (10) shows. If the size of a crystallite is sufficiently small and if the diffusion coefficient is sufficiently large, Eqs. (6), (7), (8), (14), and (15) are not valid because they are derived with the assumption of time-independent ω' values. However, it is possible to calculate the pulsed NMR signals expected when the ω' value experienced by a water molecule is not constant. The basic mathematical formalism is identical to that used in the calculations of pulsed NMR signals expected from slowly reorienting water molecules. The only difference is that B_o and θ are replaced by B and θ', respectively.

Calculations have been made for two different mathematical models of molecular motions[21,22]. In each calculation we determine the net time response for an ensemble of nuclear spin systems. Each of these systems is specified by a time-dependent θ' value. The θ' values are specified by a random spatial distribution of orientational probability symmetry axis orientations. In the rotational diffusion model, the consecutive θ' values experienced by a given spin system differ only a small amount. For the tetrahedral model, a random orientation distribution of fixed tetrahedra is assumed. Each spin system is constrained to a given tetrahedron. The θ' value undergoes random transitions among the four orientations defined by the four vectors joining the center and each of the four vertices of the tetrahedron. Quantities called correlation times are used to describe the rate of change of θ' values of the spin systems. For the rotational diffusion model the correlation time is $\tau_c = 1/6D_r$, where D_r is the rotational diffusion constant. The correlation time for the tetrahedral model is $\tau_c = (3/4)t_c$, where t_c is the average time a spin system remains in a given orientation.

The T_2^* and T_3^* values have been found from the T_2 and T_3 curves calculated for both the rotational diffusion[22] and tetrahedral[21] models. For a given B value, these theoretical T_2^* and T_3^* values are equal in the limit of rapid motion (i.e., $B\tau_c \leq 0.1$).[3] As $B\tau_c$ increases, T_3^* goes through a broad minimum at $B\tau_c$ values near unity; whereas T_2^* monotonically decreases and is nearly constant for $B\tau_c \geq 10$. Both mathematical models show this general behavior. However, for large values of $B\tau_c$ the T_3^* value in the rotational diffusion model is directly proportional to the cube root of τ_c. The tetrahedral model predicts that T_3^* is directly proportional to τ when $B\tau_c$ is large. Comparison with experimental T_2 and T_3 curves has shown that the tetrahedral model is superior to the rotational diffusion model in describing the shapes of observed transverse relaxation curves for deuterons in D_2O-montmorillonite clay gels. Consequently, the tetrahedral model is used to obtain B and τ_c values in such systems. It has also been found[16] that the tetrahedral model can be used to describe the NMR signals of the D_2O-Kelzan system.

The theoretical calculations are used as follows. A theoretical plot of T_3^*/T_2^* versus $B\tau_c$ is made. Then the experimental values of T_3^* and T_2^* are used to find the $B\tau_c$ value from this plot. A theoretical plot of $(BT_2^* - 1.457)$ versus $B\tau_c$ is then used to find the correction α for use in the equation

$$B = \frac{1.457 + \alpha}{T_2^*} \qquad (16)$$

which allows us to evaluate B from the measured value of T_2^*. The correction α is small when $B\tau_c$ is large, as shown in Table I and the value of B is insensitive to the theoretical model used in this data analysis. With decreasing $B\tau_c$ values, α increases. The value of τ_c can be found by using the value of $B\tau_c$ evaluated previously.

As mentioned in an earlier section, τ_c is related to the average time t_c a water molecule experiences the interlaminar space of one crystallite before it diffuses to a crystallite having a large difference of orientation in the magnetic field of the NMR spectrometer. This is a diffusion problem. If clay platelets are circular, the value of t_c can be estimated from the following diffusion equation[23].

$$P(t) = 4\sum_{n=1}^{\infty} \beta_n^{-2} \exp\left[-\beta_n^2 Dt/a^2\right], \qquad (17)$$

where $P(t)$ = probability of a given molecule remaining in the original crystallite at time t,
$\quad$ D $\quad$ = self-diffusion coefficient of water,
$\quad$ a $\quad$ = radius of clay platelets,
$\quad$ β_n $\quad$ = roots of the Bessel function equation $J_o(\beta_n) = 0$.

Table I.   The values of $BT_2{}^*$, $BT_3{}^*$, and α computed for
various values of BT_c with the tetrahedral model.

BT_c	$BT_2{}^*$	$BT_3{}^*$	α
1000.00	1.458	1338.6	0.001
500.00	1.459	673.11	0.002
333.33	1.459	449.94	0.002
177.80	1.460	240.88	0.003
133.33	1.461	181.48	0.004
100.00	1.462	137.20	0.005
56.20	1.465	78.616	0.008
31.623	1.471	45.022	0.014
17.780	1.482	26.428	0.025
10.000	1.502	16.031	0.045
5.620	1.538	10.010	0.081
3.1623	1.607	6.719	0.150
1.7780	1.747	4.757	0.290
1.3333	1.871	4.295	0.414
1.0000	2.061	4.052	0.604
0.7500	2.360	3.998	0.903
0.5620	2.833	4.150	1.376
0.4220	3.547	4.576	2.090
0.31623	4.594	5.373	3.137
0.17780	8.012	8.444	6.555
0.13333	10.650	10.970	9.193
0.10000	14.174	14.419	12.717

The first term in the infinite series accounts for 69.2% of the
total.   Using only this term, we obtain the relationship

$$t_c = \frac{a^2}{\beta_1{}^2 D} . \tag{18}$$

Then, since $\tau_c = 0.75\, t_c$, and $\beta_1{}^2 = 5.79$, the value of τ_c is

$$\tau_c = \frac{a^2}{7.72 D} . \tag{19}$$

EXPERIMENTAL SECTION

The observed pulsed NMR signals are affected by inhomogeneity
of the magnetic field experienced by the nuclei in the sample.
These inhomogeneities in the clay-water systems arise both from the
inhomogeneity of the electromagnet and from the magnetic properties

of some clays. Equation (2) shows that a distribution of H_o values results in a distribution of ω_o values. The effects of this distribution can be illustrated by assuming a Gaussian distribution of magnetic field values. The central magnetic field is H_o and the root mean square deviation from this value is σ. The observable transverse nuclear spin magnetization for motionless molecules is obtained by averaging the expectation values in Eqs. (6)-(8) over this distribution. After a 90° pulse, this average is

$$\langle \overline{M_+(t)} \rangle = iM_o \exp(-\gamma^2\sigma^2 t^2/2) \exp(-i\omega_o t) \cos(\omega't). \tag{20}$$

Following a 90°-180° pulse sequence,

$$\langle \overline{M_+(t)} \rangle = -iM_o \exp\left[-\gamma^2\sigma^2(t-2\tau)^2/2\right] \exp\left[-i\omega_o(t-2\tau)\right] \cos(\omega't). \tag{21}$$

Following a 90°-90° pulse sequence,

$$\langle \overline{M_+(t)} \rangle = \tfrac{1}{2}iM_o\left\{\exp(-\gamma^2\sigma^2 t^2/2) \exp(-i\omega_o t)\right.$$
$$\left. \mp \exp\left[-\gamma^2\sigma^2(t-2\tau)^2/2\right] \exp\left[-i\omega_o(t-2\tau)\right]\right\} \tag{22}$$
$$\times \cos\left[\omega'(t-2\tau)\right].$$

When the magnetic field inhomogeneity σ is sufficiently small, the transverse relaxation curve from a 90° pulse is equal to that from a 90°-180° pulse sequence. In the presence of significant magnetic field inhomogeneities, the $\cos(\omega't)$ term can be evaluated from 90°-180° sequences because the inhomogeneities do not affect the signal at $t=2\tau$. The second term in Eq. (22) also has this same dependence on magnetic field inhomogeneity. These equations are valid when the molecules are in translational motion, provided that the local magnetic field gradients, G, obey the relationship $\gamma^2 G^2 D\tau^3 \ll 1$. These considerations were taken into account in selecting the experimental conditions.

The pulsed NMR measurements were made on the deuterons at 8.0 MHz and room temperature (25°C) with conventional pulsed NMR apparatus. The T_2 curves were measured with 90° pulses when magnetic field inhomogeneity did not contribute to the signal decay. When the magnetic field inhomogeneity was significant, the T_2 curves were measured using 90°-180° pulse sequences.

The T_3 curves were measured with 90°-90° pulse sequences having a 90° r.f. phase shift between the two pulses when the effects of magnetic field inhomogeneity were insignificant. The lower, positive sign in Eqs. (8) and (22) corresponds to this pulse sequence. When magnetic field inhomogeneity effects were important, an additional magnetic field gradient was applied to the sample in order to suppress the first exponential term in Eq. (22). If this gradient had not been applied, the T_3 curve would have been in error because the relative importance of the two exponential terms in Eq. (22) would have been a function of 2τ.

The powder samples were prepared by adding the desired volume of D_2O to the weighed clay powder in 12 mm o.d. Pyrex vials. After the vials were sealed, the H_2O contents were measured by comparing the proton 90° pulse signal amplitude to that of a known amount of H_2O. Then, using the stated isotopic purity of the D_2O, the amount of dry clay and the total volume of water (both H_2O and D_2O) in the sample were calculated.

The oriented sodium hectorite samples were those which had been used in earlier experiments[5]. The other clays are described in the appropriate sections of this paper.

RESULTS

Oriented Sodium Hectorite

Measurements of the deuteron T_2 curves were made for the suite of oriented samples[5] containing 0.856 to 8.408 cm^3 of water per gram of dry clay. The shapes of the measured relaxation curves were typical of a damped cosine function. An example is given in Fig. 2. This observation indicates that all of the observable water molecules exhibit the doublet splitting. As Eq. (13) shows, if some of the water did not have a splitting, the observed relaxation curves would oscillate about some positive base line instead of zero, contrary to observations. The splitting constants B obtained by applying Eqs. (6), (7), and (10) to the measured transverse relaxation curves are given in Table II. The concentration dependence of the deuteron splitting constant is also shown in Fig. 3. Also included in this figure are the proton B values obtained[5] from a suite of samples containing 0.21 to 2.26 cm^3 of H_2O per gram of dry hectorite. These results show that the observed doublet splitting constant is directly proportional to the amount of dry clay per cubic centimeter of water.

Let us examine the implications of these results. The diffusion of the water molecules is sufficiently rapid so that during the time 1/(splitting constant) the water molecules have experienced all the different distances from the parallel clay platelets. Hence, the observed splitting is given by the average (3 $\cos^2\theta - 1$) taken over all these locations. The proportionality relationship together with the concentration range of its applicability (from some number between 2 and 6 up to 84 layers of water molecules between platelet pairs) means that the contribution of the preferential orientation of the first few molecular layers of water to the observed doublet splitting is very large compared to that for positions further

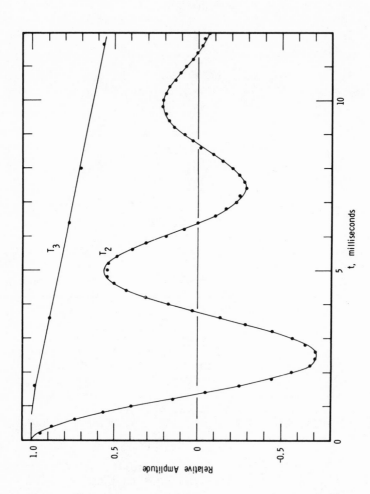

Figure 2.   The deuteron T_2 and T_3 curves obtained from measurements on an oriented sodium
hectorite sample containing 8.408 cm³ of water per gram of dry clay.   The
orientation is $\theta' = 90°$.

Table II.   Values of the deuteron doublet splitting constant,
B, obtained from samples made by adding D_2O to
oriented sodium hectorite. The concentration of
water is expressed as the ratio of the number of
cubic centimeters of water per gram of dry clay,
cm^3/g.

cm^3/g	B (sec^{-1})	10^{-4} B' (sec^{-1})
0.856	1.277×10^4	1.094
1.726	6.13×10^3	1.059
2.615	6.22×10^3	1.103
4.141	2.82×10^3	1.206
5.528	2.095×10^3	1.158
7.289	1.721×10^3	1.255
8.408	1.286×10^3	1.082

Table III   Concentration dependences of the deuteron T_2^*
and T_3^* values obtained from samples made by
mixing D_2O and powdered sodium hectorite. The
B and τ_c values were obtained by using the
tetrahedral model.

cm^3/g	T_2^* (msec)	T_3^* (msec)	B (sec^{-1})	$10^4 \tau_c$ (sec)	10^{-4} B' (sec^{-1})
1.191	0.162	1.180	9.43×10^3	6.82	1.123
2.29	0.312	1.347	5.13×10^3	6.44	1.176
4.63	0.658	2.664	2.46×10^3	12.37	1.137
9.37	1.521	4.903	1.103×10^3	20.4	1.034

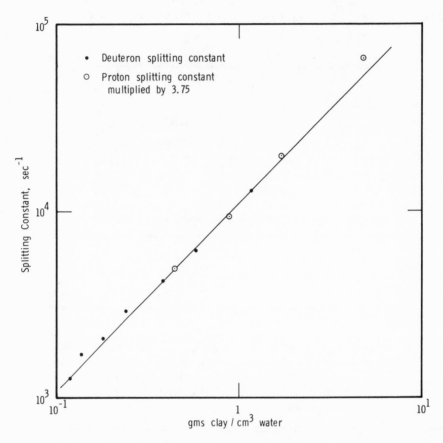

Figure 3.   The concentration dependence of the proton and deuteron
             doublet splitting constants of oriented sodium hectorite.

removed from the surface.

It is useful to define the specific splitting constant B' as the product of B in sec^{-1} and the number of cubic centimeters of water per gram of dry clay. The results given in Table II show that the value of B' is virtually concentration independent for these oriented sodium hectorite samples.

Powdered Sodium Hectorite

Experiments were performed to determine the optimum method of preparing montmorillonoid clay gels. It was found that samples prepared by stirring freeze-dried clay with D_2O followed by freezing and thawing in the sealed vial exhibited the largest B' values.

Measurements were also made to investigate the concentration dependence of the deuteron splitting for powder samples in order to determine the optimum concentration for studies on montmorillonoids. Three different sodium clays were used. The sample preparation followed the procedure given above, except that the samples with water contents less than two cubic centimeters per gram of dry clay were too stiff to allow complete stirring. Figure 4 shows the T_2 and T_3 curves measured for a hectorite sample.

The numerical results for hectorite given in Table III show that the concentration dependence of B is the same for both oriented and powdered samples. This fact indicates that the method of data analysis for powdered samples is valid. The results from all three montmorillonoids showed that the value of B' is essentially independent of concentration when the water content is less than 2.5 cm^3 per gram of dry clay. Hence, good values of B' can be obtained conveniently from samples containing 2.0-2.5 cm^3 water per gram of dry montmorillonoid.

Powdered Montmorillonoids

Values of B' and τ_c were obtained from montmorillonoid clays mined at different localities. The clays were prepared in the sodium ion-exchange form, and the clay particles with sedimentation radius less than 0.1 micron were selected by centrifugation.

The first four clays in Table IV were provided by J. C. Davidtz[*] in the sized and sodium ion-exchanged form. The Umiat

---

[*] Central Research Laboratories, Mobil Research and Development
    Corp., Princeton, New Jersey 08540.

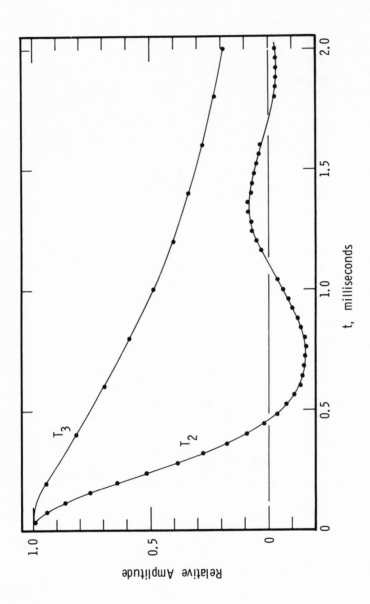

Figure 4.   The deuteron T_2 and T_3 curves obtained from measurements on a powder sample containing 2.207 cm^3 of water per gram of dry sodium hectorite.

Table IV. Values of the deuteron $T_2{}^*$, $T_3{}^*$, τ_c, and B' obtained from measurements on samples made from powdered sodium clay and D_2O. The tetrahedrel model was used to obtain the τ_c and B' values. The numbers in parentheses refer to A.P.I. Project 49 sample numbers.

Clay	cm^3/g	$T_2{}^*$ (msec)	$T_3{}^*$ (msec)	$10^4\tau_c$ (sec)	$10^{-4}B'$ (sec^{-1})
Mont., Bayard, New Mexico (30a)	2.466	0.315	0.988	4.04	1.32
Mont., Belle Fourche, South Dakota	2.318	0.325	1.143	4.97	1.18
Mont., Cheto, Arizona	2.537	0.464	0.915	2.26	1.12
Mont., Otay, California (24)	2.234	0.259	1.021	4.68	1.40
Mont., Polkville, Miss. (19)	2.360	0.283	0.860	3.45	1.42
Mont., Santa Rita, New Mexico (30)	2.288	0.242	0.894	3.98	1.55
Mont., Tatatilla, Mexico	2.176	0.264	0.848	3.52	1.39
Mont., Umiat, Alaska	2.137	0.253	1.045	4.89	1.36
Mont., Upton, Wyoming (25)	2.053	0.285	1.024	4.50	1.19
Beidellite, Black Jack Mine, Idaho	2.343	0.250	1.071	5.10	1.50
Hectorite, Hector, California	2.207	0.292	1.360	6.72	1.20
Vermiculite, Bush Deposit, Llano, Texas	2.345	0.308	4.200	27.3	1.14

clay was obtained from D. M. Anderson**. The Tatatilla clay was
provided by the U.S. National Museum in Washington, D. C., catalog
number 101836. The Polkville and Upton clays were purchased from
Wards Natural Science Establishment, Inc. The Baroid Division of
the National Lead Company provided the hectorite, which was designated B1-26. The procedure outlined above was used in preparing
the samples.

Measurements were also made on a lithium ion-exchanged sample
of vermiculite. This last clay was oriented, and the sample was
made by adding D_2O to a glass vial containing randomly oriented
1 mm. size particles which were made by cutting the oriented sheet
with a scissors. The sample was not stirred or frozen.

The results presented in Table IV show that the B' values of
the various clays are remarkably similar. The average numerical
B' value is 1.314×10^4 with a spread between extremes of only
32.2% of the average value. Figure 5 shows the T_2 and T_3 curves
measured for the Tatatilla montmorillonite.

In another set of experiments, previously prepared[24] lithium,
sodium, potassium, rubidium, and calcium ion-exchanged forms of
Belle Fourche montmorillonite were used. These samples, which had
been used in previous experiments[5], had been prepared as described
above, except that they had not been stirred. The B' values
presented in Table V shows only minor differences among the samples
containing the various cations.

Powdered Illite and Kaolinite

The preceding results showed that the splitting B' is similar
for many expandable clay minerals. It is of interest to compare
these results to those obtained from non-expandable clays such as
illite and kaolinite. The specific surface area is different for
these different types of clay minerals. Therefore, it is necessary
to normalize the splitting constants to unit surface area available
to one cm^3 of water in order to make this comparison. If 750 m^2/g
is the surface area available to water in expandable clay minerals,
the average numerical B' value for the clays in Table IV gives the
value $B'/(m^2/g) = 17.5$ for this normalized splitting.

Since the specific surface areas of illites and kaolinites are
variable, BET nitrogen surface areas were obtained for several
powdered samples. The clays and surface areas are given in Table VI.

** U.S. Army Cold Regions Research and Engineering Laboratory,
Hanover, New Hampshire 03755.

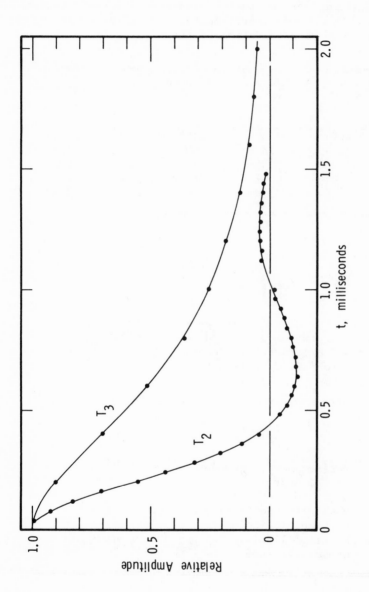

Figure 5. The deuteron T_2 and T_3 curves obtained from measurements on a powder sample containing 2.176 cm^3 of water per gram of dry sodium Tatatilla montmorillonite.

Table V.   Values of the deuteron T_2^*, T_3^* obtained from samples
           containing various cation-exchanged forms of powdered
           Belle Fourche montmorillonite. The tetrahedral
           model was used to determine the B′ and τ_c values.

Exchanged Cation	cm^3/g	T_2^* (msec)	T_3^* (msec)	$10^4\tau_c$ (sec)	$10^{-3}B'$ (sec^{-1})
Lithium	2.098	0.271	1.156	5.50	12.41
Sodium	2.039	0.326	1.443	6.98	9.99
Potassium	1.960	0.358	1.721	8.63	8.65
Rubidium	2.014	0.456	2.868	15.91	6.80
Calcium	2.054	0.389	0.902	2.84	9.84

Table VI.   Kaolinites and illites and the nitrogen B.E.T.
            surface areas measured for the samples used in
            the NMR experiments. The numbers in parentheses
            refer to A.P.I. Project 49 sample numbers.

Clay	Location	Specific Surface Area (m^2/g)
Kaolinite (1)	Murfreesboro, Arkansas	22.7
Kaolinite (4)	Macon, Georgia	37.5
Kaolinite (5)	Bath, South Caroline	14.9
Kaolinite (9)	Mesa Alta, New Mexico	17.3
Illite (35)	Fithian, Illinois	65.2
Illite (36)	Morris, Illinois	51.3
Illite	Grundy, Illinois	66.0

The clays with A.P.I. Project 49 numbers were purchased from Ward's
Natural Science Establishment, Inc. Portions of these same
powdered samples were used in making D_2O-clay mixtures. The samples
which contain approximately 0.55 cm^3 of water per gram of dry clay
were stirred. Those samples containing approximately 0.33 cm^3 of
water per gram of dry clay were too stiff to be stirred; only these
latter samples were frozen and thawed.

The results for the kaolinite samples are given in Table VII.
There is no apparent concentration dependence for the B' values.
The numerical average of the values of $B'/(m^2/g)$ in Table VII is
17.5, identical to the value for the expandable clays. The corresponding average value from Table VIII for the illite samples is
somewhat larger, 23.9. However, results from X-ray diffraction
measurements indicate some expandable layers in the illites[25].
Since these layers would not adsorb nitrogen, the N_2 BET surface
areas would be somewhat low, resulting in an increased $B'/(m^2/g)$
value.

DISCUSSION

It is pertinent to examine the implications of the τ_c values
obtained from the measurements on the samples made from powdered
clay. Values of τ_c for the montmorillonoids with sedimentation
radius less than 0.1 micron can be estimated by using Eq. (19).
The diffusion coefficient D of H_2O in montmorillonoids has been
measured[26] by neutron scattering spectroscopy. At a water content
of two cubic centimeters per gram of dry clay, the value is
$D = 1.5 \times 10^{-5}$ cm^2/sec. The substitution of this value of D and
the value $a = 10^{-5}$ cm into Eq. (19) yields the value $\tau_c = 8.6 \times 10^{-7}$
sec. This value is much shorter than those given in Table IV. A
typical τ_c value of 4.0×10^{-4} sec from Table IV corresponds to
$a = 2.15 \times 10^{-4}$ cm. Hence, the effective particle size is very
large compared to the $< 10^{-5}$ cm sedimentation radius.

The close agreement between the splitting constants obtained
from samples made with oriented and powdered hectorite (see Tables
II and III) indicates that the degree of parallelism of the clay
platelets is similar for the effective particles in the powder
samples and the oriented samples. A smaller degree of parallelism
would result in a smaller doublet splitting constant. Hence, the
effective particles in samples made from powdered montmorillonoids
are regions of near-parallel alignment of individual clay platelets.
Such regions have also been observed by other techniques and the
term "quasi-crystal" has been given[27] to them.

The values of doublet splitting constants given in Table IV
are indicative of the combined effects of preferential orientation
of water molecules induced by the clay platelets and the parallelism

Table VII. Results of the deuteron pulsed NMR measurements made on D_2O-kaolinite samples.

Clay No.	cm^3/g	T_2^* (msec)	T_3^* (msec)	$10^4 \tau_c$ (sec)	$B'/(m^2/g)$
1	0.531	3.943	5.163	5.02	20.4
4	0.542	2.425	3.843	6.27	15.2
5	0.535	7.203	8.899	6.92	19.5
9	0.525	7.994	9.462	5.96	16.8
1	0.328	2.576	3.604	4.32	17.1
4	0.336	1.454	2.305	3.77	15.7
5	0.334	4.083	5.232	4.71	19.8
9	0.322	4.694	5.822	4.61	15.4

Table VIII. Results of the deuteron pulsed NMR measurements made on D_2O-illite samples.

Clay	cm^3/g	T_2^* (msec)	T_3^* (msec)	$10^4 \tau_c$ (sec)	$B'/(m^2/g)$
35	0.549	1.542	2.048	2.09	18.3
36	0.548	0.678	1.408	3.79	31.2
Grundy	0.549	0.718	1.585	4.67	22.1
35	0.332	0.849	1.159	1.29	19.0
36	0.330	0.465	0.898	2.15	28.9
Grundy	0.338	0.403	0.921	2.85	23.9

of these platelets in the quasi-crystals. To the extent that the parallelism is similar for the different clays, these doublet splitting constants can be used to compare the preferential orientation of water molecules influenced by the different clay surfaces.

The faces of the montmorillonoid platelets consist of planes of oxygen atoms arranged in either regular or distorted hexagons. The platelet has an inner octahedral layer and two outer tetrahedral layers. Atomic substitutions in these layers result in a negative charge. Adsorbed cations preserve the overall electrical charge neutrality of the system. In octahedral substitution, the negative charge lies in the center of the platelet which is 3.30A from the plane defined by the centers of the surface oxygen atoms. The negative charges resulting from tetrahedral substitution lie in a plane which is 0.60A from this oxygen plane. The interaction between these electrostatic charges and the dipole moment of the water molecule should depend on the number of charges and on the distance between the charges and the water molecule.

Consequently, one might expect that the preferential orientation of water molecules would depend on the cation exchange capacity and on the details of the clay lattice atomic substitution. However, the relative uniformity of B' values in Table IV is an indication that these factors are of minor importance. For example, the B' values of Santa Rita montmorillonite and Black Jack beidellite are nearly equal and the cation exchange capacities are almost identical[28], but the atomic substitution in the Santa Rita clay is 98 percent octahedral while that of the Black Jack Beidellite is 100 percent tetrahedral. Similarly, the B' value of the Bayard clay is only slightly greater than that of hectorite while the cation exchange capacity[29] is 3.8 times that of hectorite. The changes in B' value with different exchanged cations given in Table V are also minor. The observed differences likely result from differences in the quasicrystalline structure[24]. These samples were not stirred. Also, X-ray diffraction studies[24] made on several of these identical clays indicate that the lattice expansion due to the water is uniform for the sodium and lithuim ion-exchanged samples while that for the potassium clay is very nonuniform.

The illite clays have tetrahedral layer atomic substitution and oxygen structure on the platelet faces, similar to the montmorillonoid clays. However, since the clay is essentially non-expandable in water, the surface available to the water has a much higher percentage of platelet edge area than is the case for montmorillonoids. The platelet edges have exposed lattice hydroxyl groups of the octahedral layer. On the other hand, the kaolinite platelet has an oxygen surface on one of its faces and a hydroxyl group surface on the other face. Hence, the nature of the surface is different for the three types of clays used in these experiments. However, the doublet splitting per unit surface area available to a

cubic centimeter of water is remarkably similar for all the clays
studied. This similarity indicates that the presence of an interface is more important than the electrical and chemical characteristics of the substrate in determining the preferential orientation
of water molecules in these systems.

It is possible to attribute the structural characteristics
of interfacial water to two influences: molecular interactions with
the substrate and interactions between water molecules. The similarity of the NMR doublet splitting constants suggests that the
latter influence is generally predominant. This may mean that the
main effect of a substrate is to provide a relatively stationary
boundary for the water molecules. Since water molecules can be
arranged in many different ways, the boundary allows the water
molecules to interact anistropically and assume an interfacial
structure which has some characteristics which are insensitive
to substrate properties. Data which suggests that the structure of
water near interfaces is different from bulk water has been reviewed[30] recently.

The preceding discussion has compared the deuteron doublet
splitting values for different planar surfaces. It is also possible
to compare the proton and deuteron doublet splittings for H_2O and
D_2O in contact with the same surface. The ratio of deuteron to
proton splitting depends on the details of the preferential orientation because the deuteron splitting is determined by the preferential orientation of the oxygen-deuteron chemical bond while the
proton splitting is determined by the preferential orientation of
the hydrogen-hydrogen vector in H_2O. This comparison has been made
for oriented hectorite[5] which has planar surfaces. Measurements of
the doublet splitting have been made for both H_2O and D_2O in
oriented samples of the fibrous material collagen [10], lithium-DNA[13],
and rayon[15]. Within experimental error, the splitting ratio is
the same in all these systems. This similarity may indicate that
there is some structural similarity of the interfacial water in all
these systems with substrates having greatly different physical and
chemical characteristics.

ACKNOWLEDGEMENTS

The author is grateful to J. T. Edwards for the preparation of
many clay samples, to G. M. Brammer for pulsed NMR measurements, and
to Dr. J. W. Whelan for the surface area measurements on the kaolinites and illites. He also thanks Mobil Research and Development
Corporation for permission to publish this work.

REFERENCES

1. A. H. Narten, J. Chem. Phys. 56, 5681 (1972).

2. P. Ducros, Bull. Soc. Franc. Mineral. Crist 83, 85 (1960).

3. A. M. Hecht, M. Dupont, and P. Ducros, Bull. Soc. Franc. Mineral. Crist. 89, 6 (1966).

4. D. E. Woessner and B. S. Snowden, Jr., J. Colloid Interface Sci. 30, 54 (1969).

5. D. E. Woessner and B. S. Snowden, Jr., J. Chem. Phys. 50, 1516 (1969).

6. A. M. Hecht and E. Geissler, J. Colloid Interface. Sci. 34, 32 (1970).

7. A. G. Brekhunets, V. V. Mank, F. D. Ovcharenko, Z. E. Suyunova, and Yu. I. Tarasevich, Teoret. i Eksp. Khim. 6, 523 (1970).

8. H. J. C. Berendsen, J. Chem. Phys. 36, 3297 (1962).

9. C. Migchelsen and H. J. C. Berendsen, in Magnetic Resonance and Relaxation (North-Holland publ. Co., Amsterdam, 1967), p. 761.

10. R. E. Dehl and C. A. J. Hoeve, J. Chem. Phys. 50, 3245 (1969).

11. J. Charvolin and P. Rigny, Compt. Rend. Acad. Sci. (Paris) B269, 224 (1969).

12. R. Blinc, K. Easwaran, J. Pirs, M. Volfan, and I. Zupancic, Phys. Rev. Letters 25, 1327 (1970).

13. C. Migchelsen, H. J. C. Berendsen, and A. Rupprecht, J. Mol. Biol. 37, 235 (1968).

14. L. J. Lynch and A. R. Haly, Kolloid-Z. 239, 581 (1970).

15. R. E. Dehl, J. Chem. Phys. 48, 831 (1968).

16. D. E. Woessner and B. S. Snowden, Jr., Ann. New York Acad. Sci. 204, 113 (1973).

17. K. D. Lawson and T. J. Flautt, J. Am. Chem. Soc. 89, 5489 (1967).

18. A. Johansson and T. Drakenberg, Mol. Cryst. and Liquid Cryst. 14, 23 (1971).

19. H. van Olphen, An Introduction to Clay Colloid Chemistry (Interscience Publishers, New York, 1963), Chap. 6.

20. A. Abragam, The Principles of Nuclear Magnetism (Oxford University Press, London, 1961), Chap. 7.

21. D. E. Woessner, B. S. Snowden, Jr., and G. H. Meyer, J. Colloid Interface Sci. 34, 43 (1970).

22. D. E. Woessner, B. S. Snowden, Jr., and G. H. Meyer, J. Chem. Phys. 51, 2968 (1969).

23. W. Yost, Diffusion in Solids, Liquids, and Gases (Academic Press, Inc., New York, 1952), p. 45.

24. W. R. Foster, J. G. Savins, and J. M. Waite, Clays and Clay Minerals 3, 296 (1955).

25. R. C. Rouston, J. A. Kittrick, and E. H. Hope, Soil Sci. 113, 167 (1972).

26. S. Olejnik and J. W. White, Nature Phys. Sci. 236, 15 (1972).

27. J. P. Quirk and L. A. G. Aylmore, Soil Sci. Soc. Amer. Proc. 35, 652 (1971).

28. L. G. Schultz, Clays and Clay Minerals 17, 115 (1969).

29. J. C. Davidtz, Ph.D. thesis, Purdue University, 1968.

30. W. Drost-Hansen, Ind. and Eng. Chem. 61, no. 11, 10 (1969).

IN VIVO AND IN VITRO SPIN-LABELING STUDIES OF POLLUTANT-HOST INTERACTION

William T. Roubal

Pioneer Research Unit, Environmental Conservation Div.,
NMFS, NOAA, 2725 Montlake Boulevard East, Seattle,
Washington 98112

INTRODUCTION

Environmental pollutants in the marine ecosystem are known to invade fish tissue. Yet, in spite of numerous studies, little quantitative data is available to delineate the nature of the deposition--the precise location of pollutant in tissues and organs involved. Even less is known of physiological/pharmacological effects (whether physical or chemical in nature--or both) which may arise with entry of pollutants.

In the well-known instance of DDT, fatty depositories constitute major receptors, yet what is known about DDT-induced alterations in metabolism in marine organisms?

Because many aspects of fish behavior are related to sensory perception of environmental stimulii, the effects of pollutants on the sensory physiology of fish is another virgin area for study; one that will require an in-depth investigation.

Nothing is known about the effects of pollutants in temperature acclimatization in fish. Are membrane lipids somehow involved, as they are in chill-sensitive and chill-resistant plants and animals? Studies aimed at elucidating these effects are under consideration at this time.

In the realm of metals, very little information is available about the impact of these on fish, and the role of proteins and lipids in metal detoxification merits attention.

In brief, the biochemical/biophysical mechanism(s) by which
pollutants affect organisms under conditions of long-term sublethal
exposure are essentially unknown. The problem is a profound one,
the questions are many, yet because of the importance of cellular
membranes in biological function, it is compelling to implicate
biomembranes in biophysical-associated changes, what and how
subtle these may be.

Laboratory controlled experiments are one thing--grossly
different from conditions existing in open seas--yet evidence
suggests that laboratory studies may approach conditions which
exist in water impounded behind dams, and at outfalls from
industries into bays and rivers. Laboratory studies should lend
new meaning to significance of impact statements concerning new
land uses and industry.

Accordingly, we are initiating host-pollutant studies by:
(1) determining sites of hydrocarbon deposition and how transfer of
pollutant occurs in vivo, (2) attempting to answer the question of
how long hydrocarbons remain in living tissue, and (3) conducting
experiments in order to ascertain effects of oil spill pollutants
in fish membrane in vitro. Concomitant with this last objective,
we are initiating studies on effects of DDT and various polychlorinated biphenyls (PCB's) on membrane. Spin-labeling is one
method under our employ.

METHODS

I. In vivo Feeding Studies Using Spin-labeled Hydrocarbons

Experimental protocol consisted of synthesizing hydrocarbons
labeled with the oxazolidine nitroxide (N-oxyl-4', 4'-dimethyl-
oxazolidine derivatives of ketoalkanes and related compounds;
referred to as doxyl-hydrocarbons; DOX-HC) (Figure 1; see Methods,
Section II for preparation), and incorporating these into fish
food.

Levels of label in food ranged from 30-100 ppm using regular
hatchery food (Oregon Moist Pellet, OMP, 1/16" dia). Since we
wanted to insure maximum food consumption by fish we first used
fingerling coho salmon (3-4 inches long) because they habitually
browse the bottom of the holding tank and consume food which sinks.
Sockeye salmon and trout which aren't so apt to feed from the
bottom were used in later studies with comparable results. Fish
were held in the Center's freshwater holding facility, and treated
food was fed to fish approximately every other day for 7-10 days
per test.

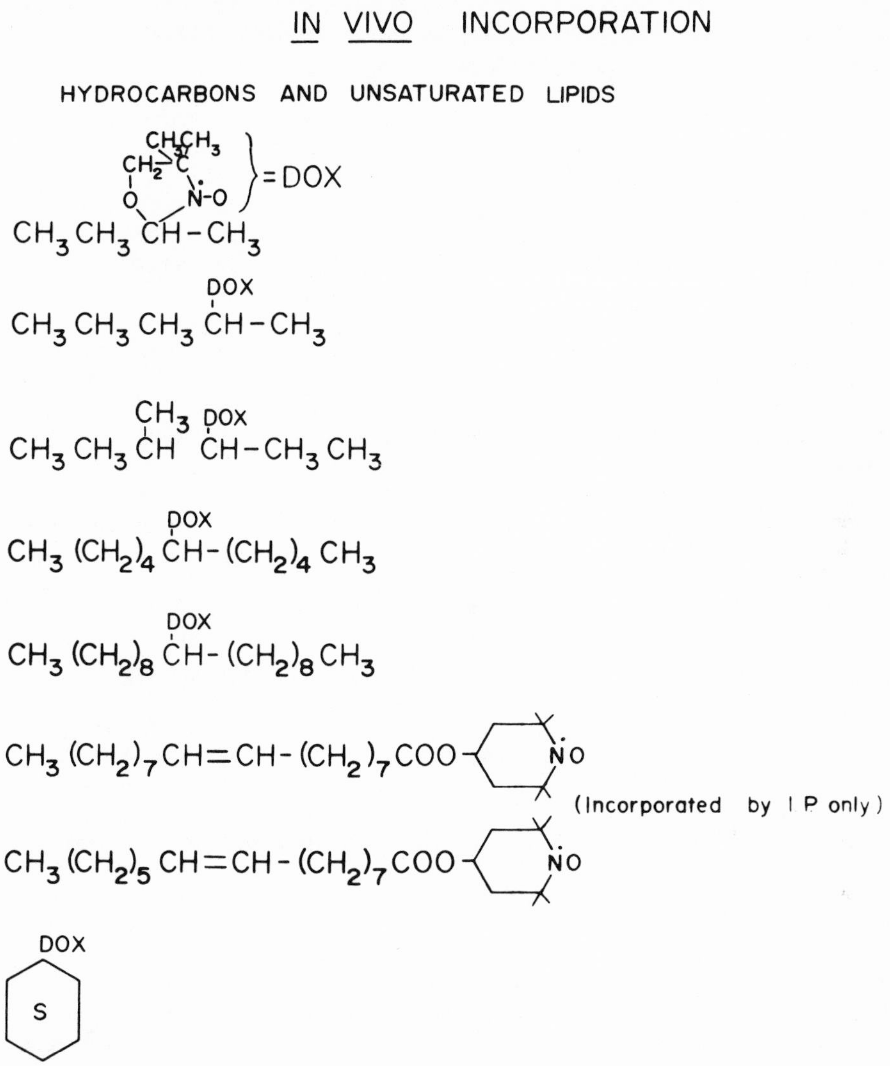

IN VIVO INCORPORATION

HYDROCARBONS AND UNSATURATED LIPIDS

Figure 1. Nitroxide-labeled hydrocarbons used in feeding studies with live fish.

The nitroxide label is part of the total oxazolidine nitroxide (DOX) and is the portion of the molecule which exhibits EPR activity. DOX derivatives are prepared as described in the text. Another class of nitroxides (tempoyl) were tested but are quickly quenched in fish. Two examples are shown here. As indicated, incorporation was only achieved by intraperitoneal (IP) injection.

II. Effects of Nonlabeled Hydrocarbons, Benzyl Alcohol, DDT and PCB's on Excised Tissue (in vitro Studies Using Fatty Acid Labels)

Because hydrocarbons appear to preferentially invade nerve tissue (see below) it was of particular interest to investigate hydrocarbon-membrane interaction using nonlabeled hydrocarbons (HC's). Here we examine the interaction of pollutants with excised tissue labeled in vitro using conventional fatty acid-type spin-labels as probes of the interaction. The reason for changing to an in vitro approach at this point in the investigation was one merely of convenience. EPR signals are more intense and an analysis of the data is facilitated.

Because spinal cord (SC) is easily removed from fish, free of bone, blood and surrounding muscle, and because spinal cord can be handled and treated with ease, only spinal cord studies are reported here. Lateral line nerve might be more desirable from a fish physiology standpoint, but this tiny bundle of nerves is excised only with extreme difficulty and does not lend itself to routine studies.

EPR measurements. An average 1" long spinal cord is about 1/16" in diameter, weighs about 15 mg (wet weight) and can be inserted by gentle suction into flared, soft glass melting point tube. The alignment and positioning is therefore maintained rather constant from test to test. Once the spinal cord has been inserted into capillary, the flare is broken off and samples in glass are merely inserted into conventional 3 mm ID quarts EPR sample tubes for study. All EPR investigations were conducted at 23°C using 1 gauss of 100 Hz field modulation and 5 mW of incident RF microwave power.

Spin labels. Four labels (Figure 2) were employed for in vitro studies. Label A is described by Hubbell et al. (1) and is synthesized in good yield with little difficulty. Labels B and C were prepared from the corresponding keto intermediate using the well-known Keana et al. (2) synthesis as reported by Waggoner et al. (3). In our laboratory, the oxazolidine intermediates leading to the nitroxide were obtained in good yield and in a relatively short time by using the "cold-contact" recycling apparatus shown in Figure 3. This apparatus maintains the water-removing efficiency of the calcium sulfate, and is a direct contrast to the usual "hot-contact" method reported in the literature (2, 3). Thus the usual 10-day reflux period of Waggoner can be shortened to 1-2 days. Aliquots of reaction mixture are removed at intervals and checked by TLC; each reaction run had a definite cutoff time beyond which oxazolidine formation

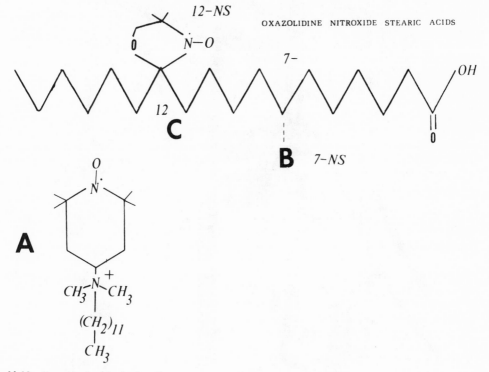

N, N–dimethyl–N–dodecyl–

N–tempoyl–NH₄ Br

Figure 2. Chemical structure of lipid-like spin-labels used for
in vitro studies of fish spinal cord membrane.

Labels B (7-nitroxide stearate; 7-NS) and C (12-nitroxide stearate;
12-NS) are doxyl-type radical labels. Label A is a tempoyl label.

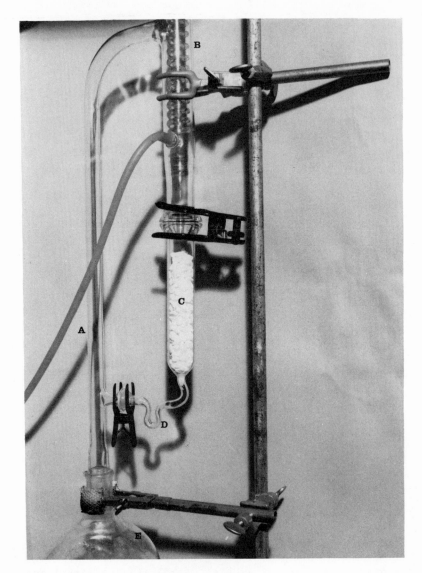

Figure 3.　Cold-contact recycling reaction apparatus for
synthesizing oxazolidines from ketones.

Hot xylene vapors containing water of reaction rise in column A,
condense in B, percolate as cooled liquid through calcium sulfate
C, and then return via trap D to the refluxing reaction mixture E.

tapers off and side reactions begin to appear. The reaction (1-2 days) is terminated at this point.

Labeling spinal cord. For the in vitro labeling of spinal cord, a few milligrams of label were solubilized in 0.25 g of bovine serum albumin in 5 ml of 0.15 M sodium chloride (isotonic). Spinal cords were added and label exchange was allowed to proceed at 4°C overnight. Labeled tissue was rinsed 35-40 minutes with several changes of fresh, cold, label-free saline, and then treated in the cold with unlabeled test compounds.

Treatment and tissue uptake. Hydrocarbon test mixtures were prepared by saturating saline with hydrocarbon and using the water-rich phase saturated with dissolved hydrocarbon for the necessary contact between hydrocarbon and spinal cord. In the case of benzyl alcohol, two concentrations were used, 20 and 100 mM, and both gave comparable results as described later.

Pollutant uptake by tissue was measured by GLC using pre-purified chloroform-methanol (2:1, v:v) for extracting C_{10} hydrocarbon and above, while freon TF (freon 113) was used for extracting aromatics and alicyclics.

DDT, biphenyl and its chlorinated derivatives were incorporated into spinal cord in the following manner: 1 mg of test compound was added to 40 mg of lecithin and the mixture was dissolved in a few ml of chloroform. The solvent was removed and 1-2 ml of isotonic saline was added. The two-phase mixture was sonicated and allowed to warm in order to disperse as much of the insoluble test substance as possible. Next, the mixture was cooled and a few milligrams of bovine serum albumin were added in order to clear most of the suspended particles. Spinal cord was added to the cooled mixture and exchange and final rinsing were conducted as described above for hydrocarbons. Spinal cords were analyzed for uptake using conventional extraction/electron capture GLC procedures.

RESULTS AND DISCUSSION

I. In vivo Uptake and Retention (Spin-labeled Hydrocarbons)

Our first interest, of course, was to ascertain if indeed nitroxide could be fed to fish and their retention measured. It was found rather quickly that this was possible.

Blood EPR activity for a variety of nitroxides (Figure 1) was observed within a short time after onset of feeding. Within one

hour, whole blood invariably exhibited a strong, three-line
spectrum showing no immobilization[1]; i.e., signals were isotropic
and quite similar to that observed for label dissolved in a low
viscosity solvent such as chloroform.

Although long chain spin-labeled hydrocarbons form immobile
complexes with blood albumin in vitro, this is not true for whole
blood removed from fish. Labels are only slightly soluble in
saline and cannot account for the intensity of signal seen in
blood. Evidently assistance from some other fluid constituent is
involved. A comparison of blood signal with labeled lipoproteins
(which give isotropic signals) leads us to believe that lipoproteins
are responsible for transfer of labels to fixed tissue sites in
fish.

When feeding of treated food was curtailed, blood EPR activity
slowly diminished, and by the end of 12-24 hours, no activity could
be detected in blood. At these later times, however, nerve tissue
exhibited activity. Additionally, EPR spectra showed some degree
of immobilization and the immobilization was greatest for the long
chain hydrocarbons. Thus the label in nerve tissue finds itself
more restricted (motionally). This is to be expected if structural
features of membrane are influential in reducing tumbling of the
label.

Spinal cord in particular, the lateral line nerve, and, to a
lesser extent, whole brain all exhibited EPR activity. Numerous
feeding trials with a wide variety of label types have shown the
doxyl type of nitroxide to be relatively resistant to metabolic
degradation in nerve tissue; tempoyl derivatives (Figure 1), on the
other hand, are quickly quenched in live fish.

Label retention appears to depend on chain length. Short
chain hydrocarbons (C_4-C_{10}) remain at detectable levels in live
tissue for 5-7 days after termination of feeding. Labeled long
chain hydrocarbons and labeled fatty acids exhibit longer periods
of observation. In the latter case, it was occasionally possible
to see signals just at detectable limits $1\frac{1}{2}$-2 weeks after food
withdrawal.

In the absence of other data, one is tempted to explain the
disappearance of EPR activity as that due to reduction of nitroxide
by living fish. Both -SH and ascorbate are effective in reducing
nitroxide, and this undoubtedly accounts for some loss of signal

---

[1] In earlier studies we reported immobilization for blood
signals (4). Recent investigations indicate this is to be true
only when labeled fatty acids are fed to fish.

in fish. Nitroxides are rather peculiar, however; on the one hand proper biological conditions cause loss of signal, yet many nitroxides nevertheless resist chemical reduction by lithium aluminum hydride in boiling benzene. One is never completely certain just what to attribute the disappearance of signal to. Preliminary studies using radioactive tracers and GLC indicate similar retention times and provide confidence in our reported retention data using spin-labels.

II. In vitro Pollutant-membrane Interaction
(Fatty Acid Spin-labels as Probes for Unlabeled Pollutants)

A complete analysis of hydrocarbon incorporation has not been made. Available data indicate levels ranging from 5-20 ppm per spinal cord for hydrocarbon, a figure comparable to data given me by colleagues studying uptake in whole flesh of intertidal mussel and the small purple shore crab (a detritus feeder) both taken from artificially polluted (laboratory studies) as well as accidentally polluted (oil spills) waters. In all cases the DDT and/or PCB uptake was 6-20 times greater than the levels given above.

EPR spectra. Representative EPR spectra are shown in Figures 4-6. In these figures, the untreated spinal cord (SC; solid line) is to be compared with treated tissue (dashed lines). Data for benzene treatment (representative of the low molecular weight aromatics), benzyl alcohol (reported in the literature as a material with local anesthetic properties (1)) and hexane (representative of a low molecular weight paraffin hydrocarbon) are provided by the various spectra. Keep in mind, however, that other aromatics behave similarly to benzene, i.e., toluene, xylene and ethyl benzene, while data paralleling hexane are observed for cyclohexane, pentane, octadecane and hexadecane.

Unpaired electron density for nitroxide resides primarily in a p-orbital on nitrogen. Additionally, in the case of labels B and C, it has been shown that this orbital lies parallel to the long axis (the amphiphilic axis) of the fatty acid chain, and hence perpendicular to the membrane surface (5, 6). These conditions are indicated in Figure 7 and account for anisotropies in the observed spectra (5-9). Less is known about preference of alignment, if any, of the p-orbital of label A. However, as will be discussed, conditions which induce immobilization (or the lack thereof) are still measurable with this label. The various contributing quantum mechanical factors which are responsible for the spectra recorded are not discussed at length in this presentation since they are discussed in the literature (6-9).

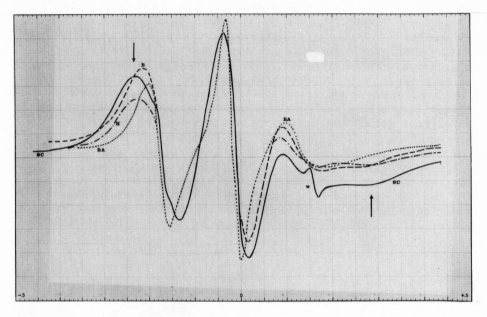

Figure 4. EPR spectra of coho salmon spinal cord labeled with
label A.

Solid line (SC) is for spinal cord alone. Treatment with benzene
(B; _ _ _ _), benzyl alcohol (BA;____) and hexane (H; __ __ _)
produce the spectral changes shown. Arrows denote anisotropic
contributions to spectra. Component w is discussed in the text.

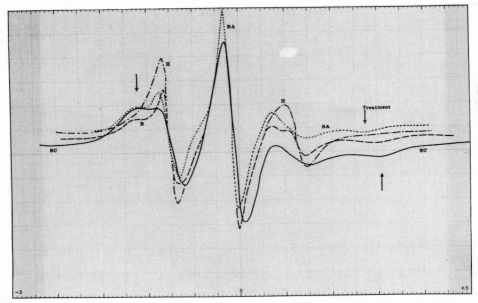

Figure 5.   EPR spectra for coho salmon spinal cord labeled with
label B.   See caption to Figure 4 for explanation of conditions.

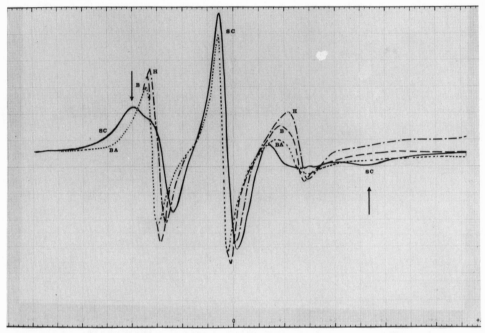

Figure 6. EPR spectra of coho salmon spinal cord labeled with
label C. See caption to Figure 4 for explanation of conditions.

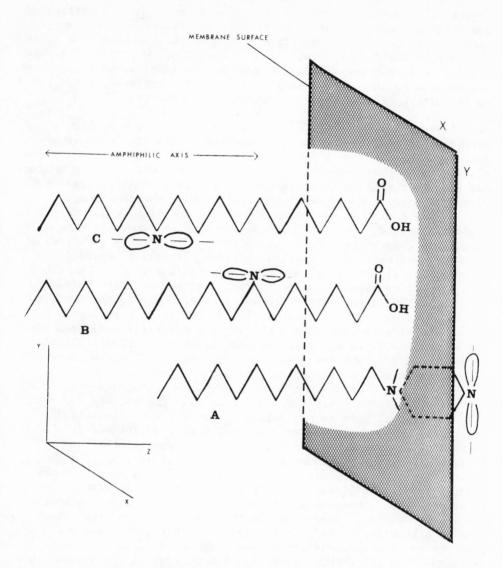

Figure 7. Schematic representation of orientation of labels A, B and C in membrane systems.

The direction of the principle p-orbital axis with respect to the amphiphilic (long axis) of the label is designated by $C\!N\!\supset$.

The spectrum of spinal cord labeled with A is shown in
Figure 4. Label A is expected to report primarily on conditions at
the membrane surface where the label is positioned due to the
influence of the label's positive charge on the quaternary nitrogen.
The sharp upfield (high field or M_{-1} component) w, typical of free
label in solution may arise from some of the label bound to membrane
but exposed to the aqueous environment. This component is reduced
on treatment. Arrows denote components of anisotropy.

Treatment with benzyl alcohol (20 or 100 mM) produces a
marked sharpening of the spectrum and immobilization is no longer
evident. Treatment with benzene, xylene, ethyl benzene and toluene
also sharpens the lines but to somewhat lesser extent, while hexane
and the other nonaromatics are least effective. Only on treatment
with nonaromatics do vestigal indications of w remain. These
results are explained by the nature of the interactions. Of the
compounds tested, benzyl alcohol may be expected to contribute most
to surface changes because of the polar nature of this compound.
Moreover, as the various spectra indicate, aromatics generally
behave similarly to benzyl alcohol. The nonaromatics on the other
hand, appear to behave as expected for systems which are unable to
form complexes of the sort which would direct them to the surface
of the membrane. In other words, the hydrophobic character of
hexane and similar compounds directs these materials away from the
surface and to sites situated more deeply into the lipid bilayer
structure of the membrane. This general trend will become more
evident as we discuss the other labels.

Data for label B are shown in Figure 5. The spectra for
untreated spinal cord is typical of a time average orientation
($2T_{11}$ anisotropy; arrows) and has been described for model
systems (5). In contrast to label A, both benzyl alcohol and
benzene affect the environment of label B the least, while the
greatest effect occurs with nonaromatics (indicated by the hexane
spectrum). In fact, anisotropic contributions remain after
treatment by both benzene and benzyl alcohol (left hand components
and arrows denoting "treatment" in the figure). Since the unique
axis of B lies in the hydrophobic interior of the membrane and away
from the polar membrane surface, the effects noted are thus
explained. Treatment with aromatics reduce some of the anisotropy
but nonaromatics are most effective in "fluidizing" this portion
of the membrane. Treatment removes anisotropy almost completely.
The data indicate that treatment results in disordering of the
membrane in regions about the nitroxide of the label.

Label C which normally only shows isotropic motion in fluid,
lipid-rich systems is illustrated by the data of Figure 6. Here
as in the case of label B we see components of anisotropy (arrows),
probably a reflection of conditions which prevail in high

cholesterol-containing myelinated tissue (spinal cord of fingerling fish contain approximately 47.5% cholesterol, 47.5% phospholipids, and 5% of lesser tissue components).

Deep within the hydrophobic interior of the membrane, the spectra indicate that all treatments are effective in disordering (fluidizing) the membrane. Careful examination of the spectra shows that nonaromatics may be a little more effective than aromatics in this respect. In fact octadecane (a solid at room temperature) fluidizes this portion of the membrane just as easily as hexane. Although myelin is expected to be more rigid than other membranes with lesser cholesterol content, it is expected that the hydrophobic hydrocarbon interior deep within the bilayer will undergo perturbations more easily than other portions of the membranes and the spectra bears this out.

Concluding this investigation, let us now turn our attention to the effects of treating fatty acid spin-labeled spinal cord with chlorinated compounds. The effects of biphenyl, chlorinated biphenyls (PCB's) and DDT on label environment are not so nicely delineated. Admittedly, the content of chlorinated compound in tissue is high, yet spectral parameters show only subtle differences between treated and untreated tissue. Differences are real, however, as they are repeatable. Figure 8 shows the nature of changes induced by both PCB's and DDT. The behavior of these labels to treatment of labeled spinal cord is provided by the data of Table I. Collectively, the data indicate slightly more mobility for label treated with biphenyl itself, followed in turn by increases in immobility with increasing chlorine content. The fact that the concentration of biphenyl derivatives range over 60-100 ppm in spinal cord immobility of label (labels A and B) with a progressive increase in chlorine content suggest that these compounds bind primarily to membrane surfaces and subsequently alter membrane organization at these locations. This is a direct contrast to treatment by unchlorinated compounds. PCB's and DDT essentially "freeze" the surface of the membrane. Observe that label C was not perturbed by treatment. In label C the nitroxide reporter group is well within the hydrophobic interstices and definitely removed from conditions which prevail at the membrane surface.

The suggestion that chlorinated compounds induce changes primarily at surfaces is in line with the data of Tinsley et al. (10) who showed (via NMR) maximum chemical shift for $-N(CH_3)_3$ protons of lecithin when DDT was added to lecithin in chloroform. Remaining choline protons underwent smaller shifts while protons of fatty acid moieties were not affected by DDT. Additional evidence along these lines has been presented in this symposium by Dr. Haque.

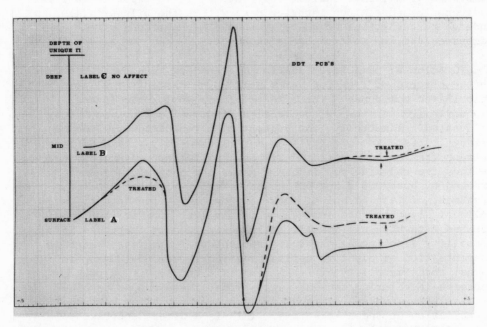

Figure 8. EPR spectra of coho salmon spinal cord labeled with labels A and B, before and after treatment with PCB's and DDT.

The slightly smaller S parameter (Table I)for biphenyl (label B) may be indicative of enhanced solubility of unchlorinated compounds in spinal cord hydrophobic regions as compared to chlorinated compounds. Although S parameter calculations are not possible for label A, the total spread between outer peaks give some indication of mobility of this label at membrane surfaces. The spread between the outer wings of the spectra climb steadily upward, ranging from 45 gauss for control to 49 gauss for one isomer of hexachlorobiphenyl. In the case of DDT this value shifts from 45 to 46.50 gauss. In addition, the ratio of M_{+1}/M_0 for DDT drops from 0.6 to 0.45 (label A), also indicating higher anisotropy for treated tissue.

Together, the noncovalent spin-labels provide a description of perturbations of myelinated tissue induced by various compounds, compounds which are insecticides or are related chemically, and compounds which are representative of oil spill pollution. The data for benzyl alcohol and nonchlorinated aromatics is consistent with other data (1) in which compounds of known anesthetic properties were tested for label perturbation in mammalian erythrocyte ghosts. Present studies show that low molecular weight aromatics alter surface organization of membrane more effectively than nonaromatics; this we believe is significant. Aromatics in general are quite toxic to marine life. Less is known about the physiological-pharmacological effects of paraffin hydrocarbons, but these appear to be considerably less toxic. Mineral oil, for instance, has been reported (11) to be nontoxic to marine mussel.

In view of recent advances in our knowledge of organization of proteins and lipids in membranes, it can be argued that changes in surface topography may be more important than changes in the hydrophobic interior (where the membrane is already the most easily perturbed region from an ordering standpoint). Topographical alterations in surface proteins and head groups of phospholipids may be principle causes of permeability changes. This appears especially true for neural processes which may involve pores or processes which exhibit pore-like properties and involve proteins or groups of proteins which communicate with both surfaces of the membrane by extending from one surface and entirely across the membrane to the opposite surface.

Membrane disorganization is considered to involve any one (or combination thereof) of at least three permutations: (a) alterations in membrane proteins, (b) changes in ion-binding properties of proteins and/or phospholipids, and (c) modifications in membrane permeability. Such changes may also alter properties of membrane-bound enzymes. Related to this is the work of Raison et al. (12) using spin-labels and Lyons et al. (13,14), who have shown the relation between molecular motion of mitochondria

Table 1. Behavior of spin-labels in spinal cord to treatment with DDT and PCB's.

Substance	Label A		Label B				Label C
	M_{+1}/M_o	Maximum separation as indicated by arrows (gauss)	T_{11} (gauss)	$T_\perp$ (gauss)	Order parameter S (see footnote)*	M_{+1}/M_o	M_{+1}/M_o
Labeled SC plus: benzene	0.6	45.00 44.50	24.8 24	10 10.1	0.6 0.57	0.4 0.62	0.4 0.69
Biphenyl		45.00	25	10.5	0.58	0.39	0.40
p-monochloro-biphenyl		47.80	25	10.5	0.58	0.39	0.40
p,p'-dichloro-biphenyl		48.50	25	10.5	0.58	0.39	0.40
3,4,3',4'-tetra-chlorobiphenyl		48.50	25.25	10.5	0.61	0.39	0.40
2,4,5,2',4',5'-hexachlorobiphenyl		48.50	25.8	10.5	0.63	0.39	0.40
2,4,6,2',4',6'-hexachlorobiphenyl		49.00	25.8	10.5	0.63	0.39	0.40
DDT	0.45	46.50			0.61	0.39	0.40

*Order parameter S indicates the degree of anisotropy. S = 1 corresponds to highly anisotropic motion, while S = 0 represents complete isotropic motion. See Seelig (7) for methods of calculation.

membrane phospholipids and enzyme activity in some plants and animals. Additionally, changes in ion-binding may occur with changes in orientation of ionic sites of membrane surfaces. Pollutants, both hydrocarbons and metals, may thus exhibit inter-relationships. Just to what extent these changes modify the physiology (and to what extent they can be tolerated) by living organisms (fish) are yet to be documented.

The picture is not complete. EPR findings must now be correlated with electrophysiological and electron microscopy studies in order to complete the overview. Studies of exposed lateral line nerve bundles in live fish look particularly promising. Just recently we have begun to explore the effects at low levels of test compounds on the electrical properties of nerve and its receptor sites in fish. We shall report on this at a later date.

Acknowledgements

The author wishes to thank Dr. Wayne J. Balkenhol for synthesizing the various spin-labeled hydrocarbons and fatty acids. Also thanks go to Tracy Collier for preparing many of the labeled spinal cords used in this investigation.

REFERENCES

1. Hubbell, W. L., J. C. Metcalfe, S. M. Metcalfe, and H. M. McConnell. Biochem. Biophys. Acta., 219, 415 (1970).

2. Keana, J. F., S. B. Keana, and D. Beetham. J. Am. Chem. Soc., 89, 3055 (1967).

3. Waggoner, A. S., T. J. Knight, S. Rottschaefer, O. H. Griffith, and A. D. Keith. Chem. Phys. Lipids, 3, 245 (1969).

4. Roubal, Wm. T. In R. T. Holman, ed., Progress in the chemistry of fats and other lipids, Vol. 13, part 2. Pergamon Press, New York and London, 1972.

5. Hubbell, W. and H. M. McConnell. Proc. Nat. Acad. Sci., U.S., 64, 20 (1969).

6. Libertini, L. J., A. S. Waggoner, P. C. Jost, and O. H. Griffith. Proc. Nat. Acad. Sci., U.S. 64, 13 (1969).

7. Seelig, J. J. Am. Chem. Soc., 92, 3881 (1970).

8. Williams, J. C., R. Mehlhorn, and A. D. Keith. Chem. Phys. Lipids, 7, 207 (1971).

9. Jost, P., L. J. Libertini, V. C. Hebert, and O. H. Griffith. J. Mol. Biol., 59, 77 (1971).

10. Tinsley, I. J., R. Haque, and D. Schmedding. Science, 174, 145 (1971).

11. Lee, R. F., R. Sauerheber, and A. A. Benson. Science, 177, 346 (1972).

12. Raison, J. K., J. M. Lyons, R. J. Mehlhorn, and A. D. Keith. J. Biol. Chem., 246, 4036 (1971).

13. Lyons, J. M. and J. K. Raison. Comp. Biochem. Physiol., 37, 405 (1970).

14. Lyons, J. M. and J. K. Raison. Plant Physiol., 45, 386 (1970).

INFRARED EMISSION FOR DETECTION AND IDENTIFICATION OF PESTICIDES

C. E. Klopfenstein and J. Koskinen

Department of Chemistry, University of Oregon

Eugene, Oregon 97403

The detection and identification of pesticides in our environment is a problem of increasing public interest. This paper describes the use of infrared emission from heated thin films of various samples to provide information essentially equivalent to that available from infrared absorbance spectroscopy.

The considerations for application of absorbance techniques to examine small amounts of material are similar to those of nuclear magnetic resonance. That is, the sample must be free of most opaque material used as filters, etc, and as concentrated in solution as possible.

We found that the emission technique gave sensitivities approaching those available in absorbance techniques, with the added advantage that opaque solid samples could be examined.

The study described below was undertaken to determine the absolute sensitivity of the instrument used and to explore various sample preparation techniques. We selected pesticide samples because of our own interest and because Low[1] has reported some work in this area.

---

(1) I. Coleman and M. J. D. Low, Spectrochimica Acta, 22, 1293 (1966).

Discussion

The considerations for optimal results in infrared emission spectroscopy are very different than those for absorbance techniques. Since the sum of transmissivity, emissivity, and reflectivity of matter is a constant, we can minimize background emission by picking a highly reflective material for the sample holder. We selected aluminum planchets which are 1.25 inches in diameter since these are reflective, convenient to handle, and about the diameter of the instrument operation. The planchets were mounted in a heating block which is shown in Figure 1.

Figure 1.   Heated Sample Cell.

Samples were prepared according to the pesticide manufacturers directions, and spread over the planchet at three times the recommended dosage. Such thin samples were required to minimize the effects of self absorbance. The reason for this can be recognized by considering that according to the emissivity relation we expect strong emission at the same wavelengths that are strongly absorbed in any sample. Therefore, in thick samples, compounds near the surface will absorb photons emitted by the lower layers at wavelengths of high emissivity, and pass photons that are weakly emitted. The overall effect is leveling, and the thicker the sample, the nearer to black body characteristics is the emission spectrum.

The total amount of pesticide used in these examples ranged from 100 micrograms to 1 milligram. Although smaller quantities could have been used, these gave reasonable intense emission signals at low temperatures, which allowed measurement times of less than two minutes for each spectrum.

Given that the number of photons per second is going to be low and limited by theory, we require a measurement technique that is as sensitive as possible. The high throughput of interferometric instruments make them much better suited than dispersive instruments for these measurements, and in fact, allow useful operation at reasonable sample temperatures without a cooled detector. We used a Digilab Model 296 interferometer for these measurements.

Experiments

The absolute sensitivity of the interferometer was determined by using a standard black body as a source, and the resulting spectrum is given in Figure 2. The shortwave emission compares favorably with the shape calculated by Plank's law. Longwave absorbance by KBr optics, etc, limits observation to above 400 cm^{-1}. In Figure 3 the emission intensity of a black body at various temperatures is compared with that predicted by Planks law. We find the fit is excellent, and stable from day to day.

The emission spectrum of polystyrene is given in Figure 4. All the strong absorbance bands strongly

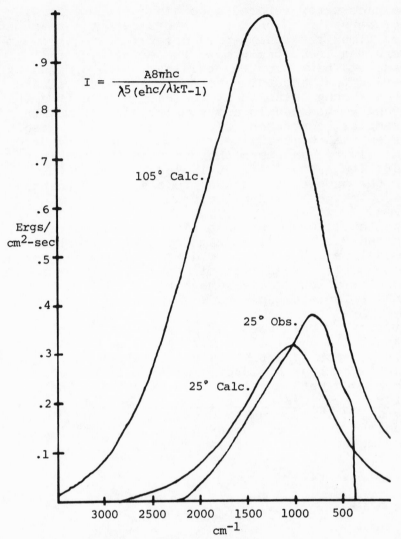

Figure 2.   Black Body Emission Detection

emit, and as expected, no measureable shift in band frequency is observed when the emission spectrum is compared to a conventional absorbance spectrum.

We next prepared samples of Diazone and Phaltan using the method described above. The absorption and

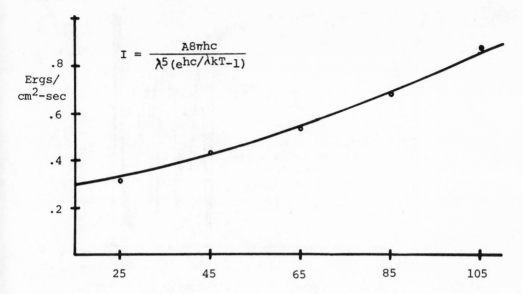

$$I = \frac{A8\pi hc}{\lambda^5 (e^{hc/\lambda kT}-1)}$$

Figure 3.   Black body emission detected at 1000 cm for various temperatures.

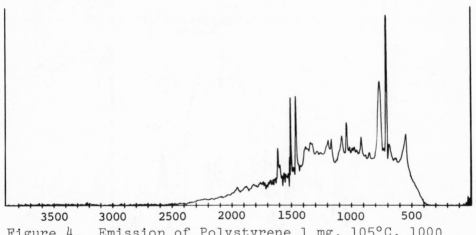

Figure 4.   Emission of Polystyrene 1 mg, 105°C, 1000 scans.

emission spectra for Diazone are given in Figures 5, 6, and 7. Quite usable data are obtained, and would allow identification of the compound without difficulty. The emission spectrum of Phaltan given in Figure 8 is also highly characteristic, and identical in line positions with its absorbance spectrum.

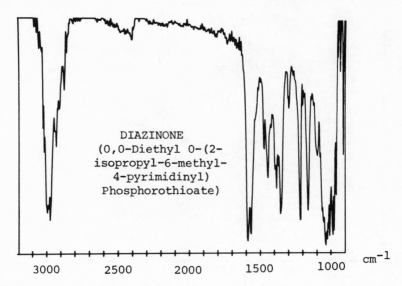

DIAZINONE
(0,0-Diethyl O-(2-
isopropyl-6-methyl-
4-pyrimidinyl)
Phosphorothioate)

Figure 5.   Diazinone Absorption Spectrum in Chloroform.

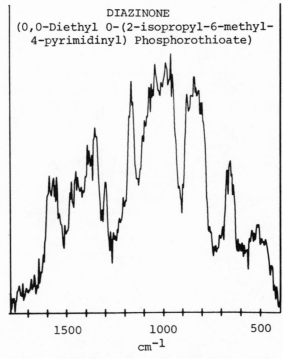

DIAZINONE
(0,0-Diethyl O-(2-isopropyl-6-methyl-
4-pyrimidinyl) Phosphorothioate)

Figure 6.   Diazinone Emission Spectrum 1 mg, 80°C, 100 scans.

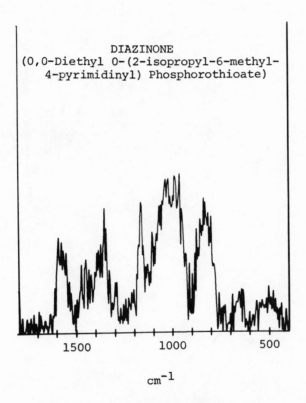

Figure 7. Diazinone Emission Spectrum 300 μg, 80°C, 100 scans.

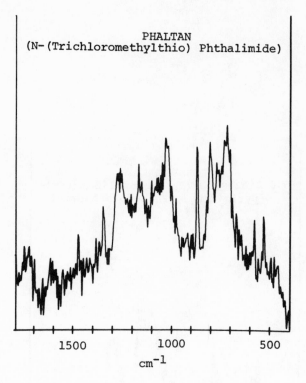

Figure 8.   Phaltan 1 mg, 85°, 100 scans

Conclusion

This study confirms the work of Low, and suggests
that infrared emission spectroscopy could be a powerful
tool for identification of pesticides, perhaps even
will still as a thin film on treated plants. Cooled
detectors for infrared detection are becomming more
readily available, and telescopic interferometers with
large light gathering capabilities could be constructed.
The combination would increase sensitivity over the
instrument used here by 1000 fold, which means that
remote sensing might be possible.

List of Contributors

C. H. M. Adams
Schood of Chemistry
University of Bristol
ENGLAND

A. L. Alford
Southeast Water Laboratory
WQO
Environmental Protection Agency
Athens, Georgia 30601

E. G. Alley
Mississippi State Chemical Laboratory
Drawer CR
Mississippi State, Mississippi 39762

Marshall Anderson
National Institute of Environmental
  Health Sciences
P.O. Box 12233
Research Triangle Park, NC 27709

F. J. Biros
U.S. Environmental Protection Agency
Primate and Pesticide Effects Laboratory
Box 490
Perrine, Florida 33157

V. W. Burse
Chamblee Toxicology Laboratory
U.S. Environmental Protection Agency
4770 Buford Highway
Chamblee, Georgia 30341

M. A. Busch
Department of Chemistry
Cornell University
Ithaca, New York 14850

D. J. Cawley
School of Chemistry
University of Bristol
ENGLAND

B. J. Corbett
National Institute of Environmental Health Sciences
National Institutes of Health
P.O. Box 12233
Research Triangle Park, North Carolina 27709

A. Curley
Chamblee Toxicology Laboratory
U.S. Environmental Protection Agency
4770 Buford Highway
Chamblee, Georgia 30341

R. C. Dougherty
Department of Chemistry
Florida State University
Tallahassee, Florida 32306

R. Haque
Department of Agricultural Chemistry and
Environmental Health Sciences Center
Oregon State University
Corvallis, Oregon 97331

O. Hutzinger
National Research Council of Canada
Atlantic Regional Laboratory
1411 Oxford Street
Halifax, Nova Scotia
CANADA

W. D. Jamieson
National Research Council of Canada
Atlantic Regional Laboratory
1411 Oxford Street
Halifax, Nova Scotia
CANADA

R. W. Jennings
Chamblee Toxicology Laboratory
U.S. Environmental Protection Agency
4770 Buford Highway
Chamblee, Georgia 30341

L. H. Keith
Southeast Water Laboratory
WQO
Environmental Protection Agency
Athens, Georgia 30601

Ute I. Klingebiel
U.S. Department of Agriculture
Agricultural Research Service
Plant Industry Station
Beltsville, Maryland 20705

C. E. Klopfenstein
Department of Chemistry
University of Oregon
Eugene, Oregon

J. Koskinen
Department of Chemistry
University of Oregon
Eugene, Oregon

Kain-Sze Kwok
Department of Chemistry
Cornell University
Ithaca, New York 14850

B. R. Layton
Mississippi State Chemical Laboratory
Drawer CR
Mississippi State, Mississippi 39762

K. Mackenzie
School of Chemistry
University of Bristol
ENGLAND

J. D. McKinney
National Institute of Environmental Health Sciences
National Institutes of Health
P.O. Box 12233
Research Triangle Park, North Carolina 27709

F. W. McLafferty
Department of Chemistry
Cornell University
Ithaca, New York 14850

B. A. Meyer
Department of Chemistry
Cornell University
Ithaca, New York

E. O. Oswald
National Institute of Environmental Health Sciences
National Institutes of Health
P.O. Box 12233
Research Triangle Park, North Carolina 27709

Stephen R. Pareles
College of Agricultural and Environmental Sciences
Department of Food Science
Rutgers University
The State University of New Jersey
New Brunswick, New Jersey 08903

S. M. dePaul Palaszek
National Institute of Environmental Health Sciences
National Institutes of Health
P.O. Box 12233
Research Triangle Park, North Carolina 27709

Gail Pesyna
Department of Chemistry
Cornell University
Ithaca, New York 14850

R. C. Platt
Department of Chemistry
Cornell University
Ithaca, New York 14850

J. R. Plimmer
U.S. Department of Agriculture
Agricultural Research Service
Plant Industry Station
Beltsville, Maryland 20705

H. A. Resing
Naval Research Laboratory
Washington, D.C. 20390

J. D. Roberts
Department of Chemistry
Florida State University
Tallahassee, Florida 32306

Joseph D. Rosen
College of Agricultural and Environmental Sciences
Department of Food Science
Rutgers University
The State University of New Jersey
New Brunswick, New Jersey 08903

R. T. Ross
U.S. Environmental Protection Agency
Primate and Pesticide Effects Laboratory
P.O. Box 490
Perrine, Florida 33157

W. T. Roubal
Northwest Fisheries Center
2725 Montlake Blvd.
Seattle, Washington 98112

F. H. A. Rummens
Department of Chemistry
University of Saskatchewan
Regina Campus
Regina, CANADA

J. F. Ryan III
U.S. Environmental Protection Agency
Primate and Pesticide Effects Laboratory
P.O. Box 490
Perrine, Florida 33157

S. Safe
National Research Council of Canada
Atlantic Regional Laboratory
1411 Oxford Street
Halifax, Nova Scotia
CANADA

Ikuo Sakai
Department of Chemistry
Cornell University
Ithaca, New York 14850

J. W. Serum
Department of Chemistry
Cornell University
Ithaca, New York 14850

K. R. Stephens
Department of Chemistry
University of Western Ontario
London, CANADA

J. B. Stothers
Department of Chemistry
University of Western Ontario
London, CANADA

C. T. Tan
Department of Chemistry
University of Western Ontario
London, CANADA

H. P. Tannenbaum
Department of Chemistry
Florida State University
Tallahassee, Florida 32306

Akira Tatematsu
Department of Chemistry
Cornell University
Ithaca, New York 14850

I. J. Tinsley
Department of Agricultural Chemistry and
Environmental Health Sciences Center
Oregon State University
Corvallis, Oregon 97331

R. Venkataraghavan
Department of Chemistry
Cornell University
Ithaca, New York 14850

E. C. Villanueva
Chamblee Toxicology Laboratory
U.S. Environmental Protection Agency
4770 Buford Highway
Chamblee, Georgia 30341

R. G. Werth
Department of Chemistry
Cornell University
Ithaca, New York 14850

N. K. Wilson
National Institute of Environmental
  Health Sciences
P.O. Box 12233
Research Triangle Park, NC 27709

D. E. Woessner
Mobil Research and Development Corp,
Dallas, Texas 75221